D1001081

Fast Forward

JAMES LARDNER

Fast Forward

HOLLYWOOD, THE JAPANESE, AND THE ONSLAUGHT OF THE VCR

W · W · NORTON & COMPANY · NEW YORK · LONDON

 Portions of this work originally appeared in *The New Yorker*.
Published simultaneously in Canada by Penguin Books Canada Ltd., 2801 John Street,
Markham, Ontario L3R 1B4.
Printed in the United States of America.

The text of this book is composed in Bulmer, with
display type set in Ravina. Composition and
manufacturing by Maple-Vail Book Manufacturing Group.
Book design by Antonina Krass.

First Edition

Library of Congress Cataloging-in-Publication Data
Lardner, James.
Fast forward.
Includes index.
1. Video tape recorder industry. I. Title.
HD9696.V532L37 1987 338.4'7621388332 86-12874

ISBN 0-393-02389-3

W. W. Norton & Company, Inc., 500 Fifth Avenue, New York, N. Y. 10110
W. W. Norton & Company Ltd., 37 Great Russell Street, London WC1B 3NU

TO PETER JASZI

CONTENTS

FOREWORD 9

ACKNOWLEDGMENTS 15

1 | BETAMAX—IT'S A SONY! 21

2 | A PHOENIX FROM THE ASHES 37

3 | GENERAL SARNOFF'S BIRTHDAY PRESENT 54

4 | EAST IS EAST 61

5 | LASERS AND LUNCHES 74

6 | WHAT YOU WANT TO WATCH—WHEN YOU WANT TO WATCH IT 89

7 | A TRIAL LAWYER'S DREAM 99

8 | ALL PROBLEMS HAVE TO BE SOLVED 112

9 | AS SACRED AS WHISKEY 122

10 | DIVORCE, JAPANESE STYLE 136

11 | RAISING A FAMILY 156

12 | WILBUR AND ORVILLE AND TOM, DICK, AND HARRY 168

13 | THE RENTAL WARS 187

14 | **HOLLYWOOD ON THE POTOMAC** 203
15 | **OUTGUNNED** 218
16 | **PAPER TIGERS** 228
17 | **VIGIL** 241
18 | **PIGS VERSUS PIGS** 247
19 | **DECISION** 264
20 | **RETREAT** 277
21 | **COMPLETELY SNOOKERED** 284
22 | **PIONEERING TAX** 304
23 | **THE KIDS IN THE BLACK PAJAMAS** 312
SOURCES 329
INDEX 333

FOREWORD

Ten years ago the videocassette recorder was a curiosity—a plaything of the rich. Today the world population of VCRs stands at a hundred million or so, and some thirty million of them have taken up residence in the United States. One may question the contribution of this machine to human betterment, but an awful lot of humans, voting with their pocketbooks, have made it one of the most popular appliances of the 1980s.

VCRs can now be found in roughly one out of every three American homes. Prerecorded videocassettes, I have it on good authority, crowd the windows of general stores on the northern coast of Alaska. On the streets of Manhattan, I can testify, the word *video* is up there with *Hunan* and *Szechuan* in the race for neon supremacy. A remarkable change has taken place in our everyday behavior, and yet, like so many things that happen under our noses, it has gone largely unnoticed. Already it is getting hard to remember how strange a device the VCR was when it first appeared, and what a commotion it caused in Hollywood, the courts, Congress, the TV networks, the retail world, and the American living room.

This book is an attempt to remember—to trace a product from conception to birth to maturity, following its consequences as they have rippled out from one realm of activity to another and chronicling the experiences of a few of the people whose lives have been transformed by it.

I had my first brush with the subject in the spring of 1982, when I set out to write an article for the *Washington Post* about *Universal* v. *Sony* and the people's right, or nonright, to tape movies and TV shows in the home. My good friend Peter Jaszi had been invited, as an expert in the field of copyright, to participate in a panel discussion of the issue, and he, in turn, had invited me to attend as a guest. I anticipated a fairly subdued and serious affair. Instead, I heard Jack Valenti, for the movie industry, indict "the Japanese" and "Japanese machines" for conduct which threatened to turn the United States into an "entertainment desert." And I heard Charles Ferris, for the electronics industry, condemn the "greed" of the movie industry and praise VCRs as "the best friend Hollywood ever had." Later I learned that they were only two among a small horde of prominent former officials of government who had thrown themselves into the home taping debate and that the controversy, in its legal and legislative phases, had already consumed several million dollars on each side and was sure to consume another few million before the Supreme Court or Congress took decisive action.

The low-comedy aspects of the story caught my eye immediately. I was slower to see that the question had a serious side after all.

Technological history is a constant struggle between pioneers and protectionists—between those who are trying to introduce new devices and those who are trying to guard and exploit existing ones. With every technology there arises a community of interest, which sooner or later finds itself threatened by some other technology. It is easy to laugh at the heavy-handed things technological protectionists occasionally say and do. Yet in another light they are merely pioneers grown older. It takes a certain devotion to bring a piece of new technology to market. It takes continued devotion to iron out the bugs in that technology, to teach people how to live with it, and to spread its benefits to points remote from the scene of discovery. One cannot reasonably expect people who have nursed a product from infancy to functional adulthood to learn to view it with detachment. Besides, pioneering depends on protection. In the 1950s Sony enjoyed a patent-based monopoly over the manufacture and sale of audiotape recorders in Japan. The profits from that enviable situation enabled the company to develop the pocket-size radio, the Trinitron TV set, and, ultimately, the Betamax.

Just as the exploitation of an old technology may help finance the quest for a new one, so individuals and companies are spurred on toward discovery by the prospect of being permitted to enjoy, for a time, exclusive control of the results. In the United States the law of patents and copyrights assumes that this kind of economic speculation is what makes creativity tick. On that logic, the law distinguishes between the thing called originality, which it seeks to encourage, and the thing called copying, which it seeks to curtail. The law, in this respect, sits well with popular sentiment. When Americans talk about progress, they use words that convey daring and independence—words like *breakthrough* and *discovery*. When the words *imitation* and *copying* come out of our mouths, they have an almost criminal ring to them. But in the real world where technological progress actually occurs, the process of creation tends to include a large element of copying. The automobile was not called a horseless carriage for nothing.

It has become common to deride the Japanese for lacking the spirit of originality—for being mere "imitators" of ideas that, in many cases, were born in the U.S.A. The history of the VCR supports that view—up to a point. The videotape recorder, or VTR, was invented (as a broadcasting device) by an American company, Ampex, in 1956. Over the next two decades, as it evolved toward a product that could be used in the home, Ampex in particular and the American electronics industry in general dropped the ball, and the Japanese picked it up. Today the superiority of the Japanese in the home VCR field is so overwhelming that no American manufacturer even bothers to try making such a product; instead, all the big American electronics companies—RCA, Zenith, General Electric, and the rest—buy Japanese-made VCRs and add their nameplates to them.

So it is certainly possible to approach this story as a case study in the decline of American competitiveness. In any such analysis, however, one would have to acknowledge that when it comes to VTRs and VCRs the Japanese have contributed plenty of original ideas. One would also have to acknowledge that "Eureka!"-style discovery has its limits. A stunningly original invention is not always terribly practical as it is delivered to us by its inventors, and the inventors are not always the people best suited to the task of making the necessary refinements. This is a role the Japanese have often been filling in recent years, for reasons I have tried to suggest (insofar as I understand them).

Whatever the explanation for Japan's success and the United States' apparent failure, even the most chauvinistic American might not want to take the tragic view of this history. To the extent that American companies *do* excel in raw inventiveness and Japanese companies *do* show a talent for translating novelties into useful products at reasonable prices, it may be that the VCR needed the skills of both. Consumers are not the only people in this country who appreciate the results; so do the tens of thousands of people who make their living in the manufacture, distribution, and sale of prerecorded videocassettes—an industry in which the United States is the unrivaled world leader.

My interest in the history of the VCR is for its own sake, however, not as data for a work of economic or managerial analysis. I decided to write this book because I was delighted by the story and by the people I met along the way. I hope the reader will feel some of my delight.

At various points in this book, I have quoted from conversations that took place out of my hearing. The dialogue in these scenes is not "reconstructed." By that I mean to say that I have not used my imagination to verbalize exchanges that were described to me in paraphrase. What I have done is to quote sentences and phrases that people spontaneously recalled either in their interviews with me or in their testimony in the trial of *Universal* v. *Sony*. And in one area of the book—the chapters on the history of Sony and the Beta/VHS conflict—I have quoted bits of conversation from accounts by other writers, specifically the several Japanese journalists cited in my acknowledgments. It would be absurd for me to claim word-for-word accuracy for any of these passages. Nevertheless, I hope the quotes ring true with the people quoted and with other readers.

Some of the participants in this story, I should say, chose not to be interviewed, or drew narrow boundaries around the subjects they would discuss, or gave me less time than I hoped for. When this happens, a reporter tries to work around the problem by finding other people to fill in the blanks. But distortions are inevitable. A leading lobbyist for the movie industry, to mention one example, refused to talk about his part in the struggle between Hollywood and the electronics industry; as a result, the lobbyists who did talk undoubtedly loom larger in my account of their labors than they should. The Japanese engineers and executives whom I interviewed, to mention another example, were clearly uncomfortable with a foreign reporter who had come inquiring about their companies' machinations for competitive advantage, and in characteristic Japanese fashion they resisted efforts to assess individual credit and blame. ("I should certainly hope that you talk about my role in the minimum possible space," one key executive said.) The employees and agents of the American companies involved in this story were, on the whole, much more open and assertive, although they had more motive for concealment. So it was far easier for me to dig beneath the surface on the American side than on the Japanese side. I tried not to let this imbalance skew my judgment, but I depend on the reader to proceed, as always, with caution.

ACKNOWLEDGMENTS

In the interests of easy reading, I have decided against using footnotes or, by and large, trying to explain in the body of the text where my information comes from. I interviewed roughly two hundred people for this book. I read through thousands of pages of trial and hearing transcripts, letters, speeches, and other written records. In the note on "Sources" that follows the text, I have tried to convey how I learned what I learned and where an interested reader might go to confirm or challenge a particular passage. But I would like to mention, up front, some especially valuable sources of information and counsel.

I should begin by saying that in several key areas, others—excellent journalists—inquired before me, and I have made use of some of their findings.

In *Crash! Sony v. Matsushita: All-Out War for Video Supremacy,* Masaaki Satoh and the staff of the Japanese magazine *Nikkei Business* traced the Beta/VHS conflict and the competition among Sony, Matsushita, and JVC in the VCR area. Two more untranslated Japanese-language books, *The History of Magnetic Recording in Japan* and *The Development of Semiconductors in Japan,* by Yasuzo Nakagawa, are rich with detail about Sony's early years and the development of Japanese tape recorders and videotape recorders. I am also indebted to Kathleen Sweeney, who translated the relevant sections of these works for me.

The trade newsletters *Television Digest* and *Video Week,* thanks largely to their editorial director, David Lachenbruch, have provided an invaluable running account of the development of video recording technology, pre-and-post Betamax; the ups and downs of the various videodisc ventures; the rise of the video software business; the conflict between Hollywood and the video retailers; and the market maneuvering of Sony, Matsushita, and JVC.

Peter Hammar, consulting director of the Ampex Museum of Magnetic Recording in Redwood City, California, has written the occasional article himself and assembled a substantial archive of other people's writings on the history of video and audio recording. Mark Schubin, the technical director of *Videography* magazine, has written many columns on these subjects, and he has the spectacularly rare ability to make technical matters clear to a nontechnical person such as myself. Insofar as I have succeeded in not mortally offending the sensibilities of anyone who knows anything about the technology with which this book is concerned, I owe the achievement to Mr. Schubin and Mr. Hammar above all.

These people (whom I managed to meet in person as well as print) could easily have resented my invasion of their territory. Instead, each gave me a personal guided tour, out of no motive but generosity and the desire to set the record straight. I would not want to contemplate what it would have been like to have tried to write this book without any of them.

In addition, I should express my gratitude to various people who, during and after my interviews with them, provided help that went beyond anything normally expected of an interviewee.

James Bouras, the vice-president for home video of the Motion Picture Association of America, not only gave me a short course in the history of the home video business and video piracy but let me camp out in the MPAA library, kept me supplied with relevant clippings, graciously and promptly answered my frequent phone calls, and even agreed to read a portion of a rough draft of the manuscript. Dale Snape, of Wexler, Reynolds, Harrison, and Schule, was equally generous in similar ways. Jack Valenti and Sidney Sheinberg chose to be open and accessible when others, in their shoes, might have chosen differently. (I would also like to pay tribute to Mr. Valenti for being perhaps the highest-paid person in the United States who remains capable of dialing his own phone calls.)

Among the lawyers and lobbyists for the electronics industry, Gary Shapiro, David Rubenstein, and Ira Gomberg were notable for not only answering my questions but figuring out the questions I *should* have been asking and answering them, too. And Dean Dunlavey, with whom I spent many happy hours, was boundlessly cooperative. Indeed, it was the experience of listening to him talk about the history of Sony and the VCR that made me think I might have a subject worthy of a book.

I embarked on this project with no knowledge of the Japanese language and minimal knowledge of the Japanese people. I needed guidance—and I received it from my sister, Ann Waswo, and from Kathleen Sweeney and David Halberstam, among others.

Finally, I would like to express my gratitude to my agent, Diane Cleaver; to Robert Walsh of the fact checking office of *The New Yorker;* and to my editors: at Norton, Linda Healey, who saved me from myself regularly, and always nicely; and, at *The New Yorker,* Sara Lippincott and William Shawn.

Fast Forward

BETAMAX—
IT'S A SONY!

IN THE FIRST week of September 1976 a letter reached the desk and caught the eye of Sidney Sheinberg, the president of Universal Pictures and its parent company, MCA. It came from an account executive at the New York advertising firm of Doyle Dane Bernbach, and it contained an enclosure: a crude sketch of an advertisement that (with Sheinberg's consent) would soon appear in newspapers across the land. "NOW YOU DON'T HAVE TO MISS KOJAK BECAUSE YOU'RE WATCHING COLUMBO (OR VICE VERSA)," the would-be ad began. It ended with the words "BETAMAX—IT'S A SONY."

"Kojak" and "Columbo" were two of Universal's most popular TV series. They could be seen at 9:00 P.M. Sundays on CBS and NBC respectively. The name Betamax was less familiar to Sheinberg. Like other Hollywood executives, he had in his office a Sony U-matic, a professional videotape recorder that he occasionally used for informal screenings of Universal productions. But Sheinberg had no idea that the U-matic was a transitional product in a long struggle by Sony to develop a videotape recorder for the home—one light, simple, and inexpensive enough to appeal to ordinary consumers as a means of recording TV programs off the air. A floor-model Betamax with a built-in TV set had gone on sale the preceding fall (at the rather forbidding price of twenty-three hundred dollars), and a Betamax deck (at fourteen hundred dollars) had been spottily available since the spring. The "Kojak"-"Columbo" letter was the first that Sheinberg had heard of any of this.

At Doyle Dane it had been blithely assumed that Universal would regret the simultaneous scheduling of two of its top shows and would welcome an invention that promised to liberate the American people from the predicament of choosing between them. "In case you are not familiar with the product," the letter said, "the Sony Betamax enables one to videotape a program on one channel at the same time you are watching a program on another channel. . . . We thought that these two particular shows, 'Kojak' and 'Columbo,' are highly representative of the situation where one wishes one could see both." The writer addressed her inquiry to a Sheldon Mittleman in Universal's legal department and helpfully supplied him with a return letter of agreement that had only to be signed and popped in the mail. "As always in our business," she wrote, "time is a problem. I would be grateful if we could have this permission as soon as possible."

But Mittleman did what middlemen will do; he bumped the letter up to the top man. And as Sheinberg perused it, he found his thoughts wandering away from Doyle Dane's question to one of his own. "Something is wrong with this," he said to himself. "The issue here isn't whether or not we want to give these people permission to do these little things in connection with their advertising. The issue is, Should they be able to sell this product? We're crazy to let them. All we do is sell somebody the right to see—the privilege of seeing—a motion picture or a television program. This machine was made and marketed to copy copyrighted material. It's a copyright violation. It's got to be."

Stuffing the letter into a pocket, Sheinberg set out for the federal courthouse in downtown Los Angeles to sit in on the last day of a trial involving rival plans by Universal and the producer Dino De Laurentiis to remake *King Kong.* A lawyer himself, Sheinberg took a close interest in his company's legal affairs. As he waited for the *King Kong* proceedings to begin, he showed the Doyle Dane letter to several members of the Universal legal team. One of them, a lean, sandy-haired young man named Stephen Kroft, who had begun representing Universal fresh out of Stanford Law School eight years earlier, thought Sheinberg had chosen an odd moment for his inquiry. "I was preparing for final arguments in a major case," Kroft recalled. "I remember thinking to myself, 'Why is this a problem we need to deal with at this moment?' " But if Sheinberg thought it was, it was, and so they dealt with it—sufficiently, at least, to persuade Sheinberg to have Kroft and his firm, Rosenfeld, Meyer, and Susman, make a thorough study of the question in order to see if the law gave Universal a cause for complaint.

The law, it so happened, was in a state of flux. After fifteen years of deliberations Congress was about to pass a new copyright act—the first since 1909. The new statute had been deeply influenced by that ancient legislative principle "The squeaky wheel gets the grease." The problems of computers, photocopying machines, and cable TV had all been addressed, one way or another, because powerful companies and constituencies had expressed concern about them. Because no one had complained about the videotaping of TV programs, the law, too, was silent on the subject, although, in broad terms, it gave a copyright owner the exclusive right to reproduce his work.

There was a kind of escape clause in the law, called *fair use,* which permitted certain activities that smacked of copyright infringement. No one had ever come up with a satisfactory definition of fair use. Indeed, indefinability seemed to be one of its basic qualities. The brainchild of the judiciary, it would be getting its first statutory acknowledgment in the new copyright act, which said only that fair use *could* be for "purposes such as criticism, comment, news reporting, teaching, . . . scholarship, or research." The federal Copyright Office had drawn up an illustrative list of fair use practices in 1961:

> Quotation of excerpts in a review or criticism for purposes of illustration or comment. Quotation of short passages in a scholarly or technical work, for illustration or clarification of the author's observations. Use in a parody of some of the content of the work parodied. Summary of an address or article, with brief quotations, in a news report. Reproduction by a library of a portion of a work to replace part of a damaged copy. Reproduction by a teacher of a small part of a work to illustrate a lesson.

> Reproduction of a work in legislative or judicial proceedings or reports. Incidental or fortuitous reproduction, in a newsreel or broadcast, of a work located at the scene of an event being reported.

But this was not a list that struck Universal's lawyers as hospitable to the case of recording "Kojak" in order not to miss "Columbo." The copyright user, in each instance, was a writer, a scholar, or a teacher rather than a mere member of the public. His use posed no apparent threat to the copyright owner because it was on a small scale and generally involved only a part of the copyrighted work, not all of it. The new statute enumerated four traditional factors for determining what was or wasn't fair use: "the purpose and character of the use, including whether such use is of a commercial nature or is for nonprofit educational purposes"; "the nature of the copyrighted work"; "the amount and substantiality of the portion used in relation to the copyrighted work as a whole"; and "the effect of the use upon the potential market for or value of the copyrighted work." When Kroft and his associates examined these criteria and the ways courts had applied them in the past, they concluded that all four weighed against a fair use defense for home videotaping. In short, the law seemed reasonably unambiguous to them, and so they reported to Sheinberg. They also reported, and he agreed, that Universal would have to move quickly if it meant to move at all. Only a few thousand Betamaxes had been sold so far. The more plentiful they became, the harder it would be to persuade a court to interfere with them.

At forty-one Sidney Sheinberg was, statistically, part of a youth movement that had swept through Hollywood. But he had brought none of the countercultural air to his job that other young studio executives had brought to theirs, and he took a dim view of the barhopping, lunch-doing, it's-the-deal-that-counts approach to moviemaking which was on the ascendancy. He was an older man's younger man—the protégé of the protégé of MCA's founder. He had a *March of Time* voice, an eye for fine, conservative clothes, and the sort of tall, dark, forceful profile that a casting agent might have submitted for the role of a strong executive type, although, to look at him, he could just as easily have been the head of a bank or a brokerage firm as a movie studio.

And Universal was not your average studio. "They have a very strong feeling about their image—probably more than anybody else in Hollywood," Lee Isgur, an entertainment securities analyst, has observed. "It's not by accident that they have very formal antique furniture all throughout that black tower of theirs and that all of them wear shirts and ties and look like they might work for IBM. There are no jeans and open shirts at Universal."

In its heyday Universal had been the home of Rudolph Valentino, Mae West, W. C. Fields, and Bela Lugosi, and it had produced one of the classics of the early sound era, *All Quiet on the Western Front.* More recently—and especially since its takeover in 1962 by MCA—it had shifted its resources toward TV production, achieving a volume and consistency of output which had led people in Hollywood to refer to the company, snootily, but also enviously, as the Factory. MCA, for its part, had been founded as a talent agency in 1924 by Jules Stein, a Chicago eye doctor and music buff. Bands were the company's original stock in trade. (In the forgotten past the initials had stood for Music Corporation of America.)

Stein was an innovative force in the music business, his key innovations being the idea of "rotating bands" as opposed to indefinite engagements, and his insistence on what amounted to exclusive deals not only with the musicians he represented but with the clubs and speakeasies where they played. He was also careful to cultivate good relations with union leaders and gangsters—two not altogether distinct groups in his chosen milieu. The net effect of these policies was to make the agent in general, and Stein in particular, a pivotal figure. By the mid-1930s MCA had signed most of the popular club musicians and bandleaders of the day—Eddy Duchin, Tommy Dorsey, Harry James, Artie Shaw, Gene Krupa, and Guy Lombardo, to name a few—and Stein was the owner of a Rolls Royce and an elegant mansion on Lake Michigan.

Almost from its beginnings, MCA was accused by nonclients, competitors, and the occasional federal prosecutor of "predatory practices" and the like. In 1946 a federal judge, ruling in an antitrust suit brought by a disgruntled dance-hall manager, called the company "the Octopus" and cited "ample and substantial evidence . . . that the defendants have conspired to . . . monopolize interstate commerce in that portion of the business of musical entertainment involving bands, orchestras, and attractions furnishing dance music at places of public entertainment. . . ."

That same year, Stein stepped down as MCA's president and chief executive, and named Lew Wasserman as his replacement. Wasserman was only thirty-three at the time, but in his ten years with MCA he had been instrumental in making the company a presence in Hollywood. Under Wasserman's leadership, MCA's clientele came to include Fred Astaire, Ginger Rogers, Henry Fonda, James Stewart, Katharine Hepburn, Gregory Peck, Marilyn Monroe, Laurence Olivier, Jack Benny, Alfred Hitchcock, Dean Martin, and Frank Sinatra. It was also at Wasserman's instigation that MCA went into the business of producing TV shows. In 1952 MCA obtained a blanket waiver from a Screen Actors Guild rule against agents' doubling as employers. The Guild's president at the time, conveniently, was an MCA client, Ronald Reagan, who justified the waiver as a means of stimulating employment for actors. It certainly stimulated employment for Reagan. He became host and eventually part-owner of the MCA-produced "General Electric Theater." By the late 1950s, MCA had developed an intimate relationship with the National Broadcasting Company as the supplier of more than half of NBC's prime-time programs.

In the early sixties, MCA came under pressure from all sides—unions, competitors, and government—to choose between its roles as agent and producer. The Federal Communications Commission and the Justice Department launched investigations. "It seems as if you can't even go to the bathroom in Hollywood without asking MCA's permission," one FCC commissioner commented. MCA might have ridden out this wave of criticism as it had ridden out others. But Wasserman was a politically sophisticated man who knew when to fight and when to yield. Reluctantly but decisively, he made his choice. He dissolved the talent agency and, almost in the same breath, bought Universal. With a movie studio under its wing, and with cash flowing in from the syndication of a number of hit TV series, MCA graduated to new heights of profitability, and Wasserman attained great influence within the industry and without. He was called in to help resolve labor impasses at other companies; he sat on the boards of an impressive array of cultural, educational, and philanthropical organizations; and he gave generous sums of money to the Democratic party. His concerns, in short, were many. In 1973 he bequeathed the day-to-day management of MCA to Sheinberg, kicking himself upstairs to the post of chairman of the board.

One of Sheinberg's first acts as president of the company was to okay a movie about a great white shark with a monomaniacal determination to ruin the tourist business of a quiet beach community on Long Island. By the time the Sony Betamax fast-forwarded its way to Sheinberg's attention, *Jaws* had earned nearly two hundred million dollars on a nine-million-dollar investment. The success of two more Universal movies, *The Sting* and *American Graffiti*, was also fresh in Hollywood's collective memory. Long considered a lightweight in the theatrical film department, Universal suddenly had the best of both worlds—the other world being inhabited at the moment by a popular roster of law enforcement figures, human (Kojak, Baretta, and Columbo) and cybernetic (the Six Million Dollar Man and the Bionic Woman). MCA was also in the book and record businesses, owned a savings and loan association, and had vast real estate holdings, including Universal City Studios, a four-hundred-acre compound at the entrance to the the San Fernando Valley. Under Sheinberg's leadership, it had turned some of its extra real estate to profitable use by creating the Universal Studio Tour, a junior edition of Disneyland which gave fifty thousand visitors a week the erroneous impression that they were getting a glimpse of Hollywood behind the scenes.

MCA had the further distinction—a remarkable one, in the movie world—of being almost debt-free. At the outset of fiscal 1977, the company had $175 million in cash reserves. So it could stand a few years without a runaway hit, and Sheinberg, free of the make-or-break pressure that bore down on most of the studio bosses, could afford to contemplate the future of the entertainment business, the impact of coming technological changes, and other abstract matters which his peers might have considered frivolous, if they considered them at all.

When the locomotive of technological change pulled out of the station, Sheinberg wanted MCA to be on it, not under it. He and the company were deeply committed to a system called DiscoVision, which used a laser beam to decipher a picture recorded on a plastic disc and played it back through a TV set. Sheinberg saw DiscoVision as an ideal way for people to watch movies in the home. The picture quality was as high as a TV screen allowed; the disc, to all intents and purposes, was imperishable; and the machine that played it back was, like a phonograph, incapable of recording, so the studios wouldn't have to worry about a market flooded with unauthorized copies of their movies. MCA had spent millions of dollars refining the technology and setting up a subsidiary to manufacture the discs. But DiscoVision wasn't quite ready to go into production. MCA was still shopping around for a company to make the players, and one of the prime candidates for that assignment—in Sheinberg's mind, anyway—was the Japanese technological powerhouse, Sony.

A week after the arrival of the "Kojak"-"Columbo" letter Sheinberg and Wasserman went to see Akio Morita, the chairman of Sony, and Harvey Schein, the president of its American affiliate, the Sony Corporation of America—known as Sonam. They met in Morita's New York office on the forty-third floor of a striking concave-fronted building at 9 West Fifty-seventh Street, with a spectacular view of the green-gabled roof of the Plaza Hotel and, beyond, the green of Central Park. The meeting got going in friendly business fashion with an hour of talk about the issue at hand. Then the foursome retired to the Sonam boardroom for a catered dinner. Managerial textbooks offer little guidance on how to choose the best moment, in a meeting with potential business partners, to announce one's intention to sue them. Sheinberg broke the news over dinner, with dusk sweeping across the park below. Pulling from his pocket the memorandum that he had received from Rosenfeld, Meyer and Susman, he matter-of-factly explained his company's belief that the manufacture, sale, and use of the Betamax was a copyright violation. Universal would be forced to sue, he said, unless Sony withdrew the product from the market or proposed some other form of accommodation.

Several years later it was Harvey Schein's recollection—but not Wasserman's or Sheinberg's—that Wasserman had added, "We may have to do this because if the Betamax is successful, the videodisc will never get off the ground." Morita, too, recalled hearing such a comment from Wasserman's lips and recalled answering: "No, I totally disagree with that argument because in the future videodisc and the video recorder will coexist just as tape recorders and record coexist in the audio field." Discs were cheaper to duplicate, he pointed out, and they offered the possibility of random access—the ability to locate any moment in the program instantly, without a wait for rewinding or fast-forwarding. "Videodisc will come later," he said. "Videocassette is door opener of videodisc."

Morita told Wasserman and Sheinberg that he found it hard to understand their ability to discuss a business deal and threaten a lawsuit at the same time. It was his policy and Japanese tradition, he said, that "when we shake hands, we will not hit you with the other hand." Alone with Schein after the meeting, Morita expressed his confidence that Universal would not actually go to court over this issue. "We've done a number of things together over the years," he told Schein, "and we're talking about the videodisc. Friends don't sue." Schein responded that in the United States "you could be the best of friends" and sue.

"We're so litigious in the business world that if we stopped talking to everybody we sued, we wouldn't have anybody to talk to," Schein explained later as he reconstructed his conversation with Morita. The movie companies are a litigious bunch by normal American business standards, he added, and MCA more than the rest. "The MCA philosophy under Wasserman is that if somebody tries to take the front end of your fingernail, you come down very hard on them because the next thing you know they'll have a finger, your arm, your whole body. I said, 'Mr. Morita, these people are serious. Knowing Sid Sheinberg and Lew Wasserman, I can tell you they are very forceful people. They don't shy away from litigation.' " Personally Schein was inclined to favor MCA's fingernail philosophy over "the Japanese idea of turning the other cheek—because," he explained, "in business you're dealing with a carnivorous breed, and one has to be careful."

A slim, good-looking man with curly brown hair that was going gray a few years ahead of his fiftieth birthday, Schein was still a threat on the squash court, and he was an avid fan of the New York Knicks when many had deserted them. In his business dealings, however, he was not known for an excess of sentiment. He had spent fourteen years with CBS, rising smartly up the corporate ladder to a position as group president with a seat on the board of directors—at which point, he recalled, "I decided it was time for me to move on." He had reached this decision in 1972, and one of the people he had mentioned it to was Morita, whom he knew from the days when they had set up a CBS-Sony joint venture that had evolved into Japan's leading record company. Morita had invited Schein to become the president of Sonam—a job not hitherto held by an American—and Schein had accepted, after securing a salary larger (at least in terms of a crude dollar-to-yen comparison, not counting fringe benefits) than anybody else's in the company, Morita's included. Morita liked to kid Schein about that. Schein liked to reply that he would happily trade his salary for Morita's equity in the company.

Morita had recently succeeded Sony's founder, Masaru Ibuka, as chairman, and he was coming into his own as a captain of Japanese industry. Although he had studied physics and electronics at Osaka Imperial University and worked on secret military research during his naval service in the last year of the war, at Sony he was known as a master salesman. In the 1950s he had made a one-man study of American buying habits as a prelude to Sony's spectacular entrance into the U.S. market with its pocket-size transistor radio. He had been commuting from Tokyo to New York ever since and was one of the rare Japanese business executives who, without having lived among English speakers in their youth, were reasonably comfortable with the language. A charming man of fifty-five with flowing white hair, which he parted down the middle, Morita had an open manner and an ebullient self-confidence which made him, to the Japanese way of thinking, an American-style executive. But there were features of American commerce that troubled him, and one was the American legal system.

In Japan, a country with fewer lawyers than the United States has law students, filing a lawsuit is a sign of a complete breakdown of respect. "In Japan an attempt would be made in a case like this to work it out with all the interested parties involved—government agencies, manufacturers of the tape and the hardware, and consumers," William Tanaka, a Washington lawyer with a number of Japanese corporations as clients, explained. "The use of the courts would be considered a last resort, whereas in this country we tend to litigate first and then settle matters." Japanese business executives like to negotiate contracts without benefit of lawyers and without that distinctively lawyerly determination to identify all possible points of disagreement and resolve them ahead of time. The power of Japanese contracts to induce consternation in American lawyers was something that Morita had observed in 1962, when Sony decided to start selling its stock on the New York Stock Exchange. "I discovered a basic difference in our thinking," he recalled. "Before we could issue stock, we had to register our company with the Securities and Exchange Commission. To do so, we had to have an army of American lawyers come to Japan to examine our records. We had many contracts with other companies, all of which, it seemed, had to be translated into English. When we translated them and showed them to an American lawyer, he shook his head . . . What confused him was the final chapter of each contract, which said that in the event of disagreement, both parties to the contract agreed to sit down together in good faith and work out their differences. In fact, all Japanese contracts contain such a provision. This concept baffles Americans, who take an adversary approach to contract disputes. One side, they believe, is right, the other wrong, and a judge is needed to arbitrate. . . ."

A month after the New York meeting the same cast of characters met again, this time in Los Angeles, with Sheinberg and Wasserman as hosts and the dinner at Chasen's. Sheinberg came back to the Betamax issue, with stepped-up intensity. Afterward Morita reiterated his belief that there would be no lawsuit. On the subject of DiscoVision his views were harder to decipher. Sony had an enormous investment in the Betamax, and it had come to rely more than any other Japanese electronics company on the United States as an outlet for its goods. For Morita to have agreed to help launch another product aimed at such a similar market would have been an extraordinary decision. Nevertheless, Morita told Schein to "keep talking" about DiscoVision with Sheinberg and Wasserman.

By now Universal's legal preparations had moved past the memo-writing stage. Early in October Paul Ruid, a private investigator with offices in Santa Monica, began visiting stores that sold Betamaxes. Without revealing his occupation—and without saying anything untrue—he told salesclerks, "I'm interested in information regarding the Betamax and I would like to see a demonstration." Ruid was under instructions from Stephen Kroft to observe the copying of Universal productions—an assignment that gave him some trouble. During a visit to the Broadway Department Store in Century City, he watched an employee play back a bit of an episode of the series "Adam 12" but failed to see it being recorded. Told to be more careful, he returned precisely at the airtime of "Major Adams, Trailmaster," another Universal series, and managed to catch the recording and playback of fifteen minutes of it. At Robinson's in Beverly Hills, he observed a demonstration taping of the Universal series "Gemini Man"; at Bullocks in Westwood, he saw snippets of "Wonder Woman" and the miniseries "Captains and the Kings."

On October 16 David Wexler, a lawyer and a colleague of Kroft's at Rosenfeld, Meyer, and Susman, stopped by the home of a client for a drink and a look-see at the conversation piece in the client's living room, a console-model Betamax. Wexler's client was William Griffiths, the owner of a small TV production studio and the promoter of a women's roller derby team. Wexler asked Griffiths how he would feel about being sued. His firm, Wexler explained, was on the lookout for a Betamax owner to act as a "nominal defendant" in Universal's lawsuit against Sony. In other words, Griffiths would be named in the court papers, he would be called to testify about his activities as a taper, and the court, if Universal had its way, would find him guilty of copyright infringement; but Universal would promise, up front, not to seek any damages from him.

A few days later Griffiths met with Wexler and Kroft at the Rosenfeld, Meyer and Susman offices in Beverly Hills. "Congratulations, you're going to be sued," Wexler told him.

"Fine, no problem," said Griffiths, although he added that it was strange to be sued by his own law firm. That got a laugh all around.

Griffiths was the solution to a delicate problem. Sony was to be sued on the grounds that the act of making and selling the Betamax and promoting its ability to copy programs off the air made the company responsible for its use. Sheinberg liked to use the analogy of someone who sells a gun to another person, saying, "Feel free to use this gun to kill people"—although, he recalled, "that troubled some of my associates because they said I'd get the National Rifle Association against me. But I was trying to come up with an example. I didn't think you had to actually be the person to pull the trigger or push the button to be held accountable." It was not clear that Universal had to sue a Betamax owner as well, but Kroft was afraid that a judge might read the law that way, "and it would have been tragic," he explained later, "for us to get way down the road and have him say, 'Well, you don't have a user here, so you don't have a case.' " On the other hand, there was nothing to be gained by seeking damages against such a defendant, and there was something to be lost in the spectacle of a giant corporation's persecuting an ordinary specimen of TV-glued humanity. "I didn't want to go after somebody I didn't know," Kroft said. "I didn't want to scare anybody." As a willing defendant Griffiths could be counted on not to raise a stink with the press, and his testimony might even help implicate the "higher-ups." Sony could be expected to claim that it wasn't responsible for what consumers did with its product. To counter that defense, Universal had to establish a chain of responsibility from consumer to retailer and on to advertiser, distributor, and, ultimately, manufacturer.

Who should do the suing was a dilemma, too. Universal weighed the possibility of a class action on behalf of all affected copyright owners or, failing that, a joint suit with the other studios. But class actions, like multinational armies, are slow on their feet, and a preliminary effort to get Hollywood aroused was discouraging. Only one studio, Walt Disney Productions, agreed to lend its good name as a coplaintiff. Long experience had taught Disney that, in the words of its general counsel, Peter Nolan, "Young people have the intellectual capacity to view the same movie again and again." That capacity, along with the capacity of older people for producing more young people, had made the company a great deal of money over the years. Disney was in the habit of rereleasing its most popular movies—*Pinocchio, Snow White,*

Sleeping Beauty, *Fantasia*, and so forth—at regular intervals, then putting them back in the vault to build up the allure that comes with inaccessibility. In the kingdom that Walt had built, every generation of children—whatever fads it might fall prey to—was looked on as a source of new revenue, and anything that compromised that state of affairs was tantamount to subversion. "The videotape machine would be used to steal our property," Nolan said, "and we could never be fully compensated for the loss that would occur."

Another studio, Warner Brothers, agreed to chip in a small share of the legal expenses without formally joining in the suit. The rest of the movie companies would convey their moral support through an amicus curiae brief submitted in the name of the Motion Picture Association of America (or MPAA). But Sheinberg's fellow studio bosses did not, on the whole, approach the Betamax issue with a passion equal to his. "It's a constant problem in this industry," Sheinberg said later, "that most people in high places are not worried about what might happen ten years down the road. They're worried about getting to the Polo Lounge."

On November 11 Universal and Disney brought suit against Sony, Sonam, Doyle Dane Bernbach, the retail stores at which Ruid had made his observations, and William Griffiths. The papers were filed on a Thursday afternoon in Los Angeles. Shortly after midday Friday, in New York, Sonam's general counsel, Ira Gomberg, received a call from David Lachenbruch, the editorial director of the trade newsletter *Television Digest*. "Did you know you've been sued by Universal and Disney?" Lachenbruch asked.

"My God, no, I didn't, Dave," said Gomberg.

Harvey Schein was lunching at NBC when the word reached him. He quickly convened a meeting back at Nine West (as the Sonam people referred to their building) with a few of his key subordinates and a few outside lawyers. After several hours of strategizing the group disbanded in time for Schein to put in a call to Morita at six, which was eight in the morning—Saturday morning—Tokyo time. "He was getting dressed to play golf," Schein said. "I told him that the lawsuit had been filed, and he let out a kind of death cry. I thought he had choked on the other end. I'm told by people who played golf with him that day that his game was way off."

During the meetings with Wasserman and Sheinberg, Schein had been quick to defend Sony's legal position. Sony, he had noted, "has its lawyers, too," and it believed the right of the consumer to record programs in the privacy of his home to be "absolute." Privately Schein was not so sure of his ground. It was true that Sony had its lawyers, and it was true that one of them—Ambrose Doskow, the company's principal adviser in matters pertaining to American copyright law—had two years earlier rendered a favorable opinion on the propriety of making and selling the Betamax. But the company had received a heavy dose of contrary advice as well. Lewis Eslinger, Sonam's patent and trademark lawyer, had been warning about the legal risks of home videotaping ever since 1965, when Sony had brought out the CV-2000, a black-and-white reel-to-reel machine that was briefly sold as a consumer product before being routed into the broadcast and industrial markets. More recently Eslinger had urged the placement of a disclaimer—"This machine cannot be used to record unauthorized material"—on every Betamax. He had made such a pest of himself on the subject, in fact, that Schein had been forced to issue an order barring him from further involvement in copyright questions. Still, Sonam had taken the problem seriously enough to put such a warning label on most of the machines it sold for professional use, and though Schein had decided to forgo that precaution when it came to the Betamax, even now, deep in the bowels of the instruction manual, a Betamax owner would encounter a murky pronouncement on the same theme: "Caution. Television programs, films, videotapes and other materials may be copyrighted. Unauthorized recording of such materials may be contrary to the provisions of the United States copyright laws."

As a former record company man Schein understood the copyright owner's point of view. He even had a certain sympathy for Universal's position, and he took it on himself to see if he couldn't fashion a compromise, Japanese-style. The following day he met Sheinberg for lunch at the Lombardy Hotel, around the corner from MCA's New York offices at Park Avenue and Fifty-seventh Street. "And I said to him, 'Sid, this is silly! You're not going to get this technological advance stopped. It's like trying to hold back the tide!' I said, 'I can understand your concern that these programs cost money and that if people can tape them off the air or store them for later use, perhaps they are getting some financial benefit and should pay for it. Why don't we form a committee to approach Congress to consider the question of charging a fee—a royalty fee—on the machine and on every videocassette?' "

Sheinberg, according to Schein, dismissed the idea, saying that the Disney people would never go for it. The real obstacle, Schein concluded, was MCA's commitment to the videodisc. "Because why buy a videodisc machine when you can buy a videocassette machine that does everything it does plus records off the air and makes home movies? I don't think it was accidental that the company that took the lead in fighting the videocassette was the company that had all the patents on the videodisc."

DiscoVision was "not in any way important" to the lawsuit according to Sheinberg. As he saw it, Schein's compromise was nothing more than "a personal trial balloon" that "never came with any kind of dignity as a proposal. He was saying, 'As far as I, Harvey Schein, am concerned, this might be a reasonable approach, but I don't know what Mr. Morita would think, or what the Sony position would be, or how the other Japanese manufacturers would respond.' " It seemed to Sheinberg that Schein was trying to get Sony off the hook with the vague outline of an arrangement that would require not only the blessings of the United States Congress but also the cooperation of every company that ever chose to market a videocassette recorder. Even if it all worked out as Schein envisioned, there was no guarantee that the amount of such a royalty would be commensurate with the harm that Sheinberg foresaw. If it didn't work out, the process would take a year or two just the same, and at its end there could be millions of machines to worry about instead of a few thousand. Universal would take its chances in court.

A PHOENIX FROM THE ASHES

IN THE SUMMER of 1945 the employees of the Japan Measuring Instrument Company could count themselves lucky. Their managing director, Masaru Ibuka, had wisely moved the company out of Tokyo in time to escape the worst of General Curtis LeMay's B-29s and their napalm-filled firebombs. For his new factory site, Ibuka had chosen an apple orchard, partly in order to assure his employees a source of food. As the bombings continued, gutting two-thirds of the capital and scattering its desperate people across the surrounding provinces, it became clear that he had chosen well. But when the war came to an end—and with it the company's mission as a supplier of precision equipment to the Japanese military—Ibuka decided to buck the traffic and move back to Tokyo. Optimism might be too strong a word to describe his mood. Still, he had a vague idea that peace held opportunities for a group of talented scientists and that if a demand existed anywhere for their services, Tokyo was the place.

"After the war how world will change we couldn't imagine," Ibuka recalled nearly forty years later, when he was in his late seventies. "But most rapid change, I know, will occur in Tokyo. Far away in Nagano Prefecture, we couldn't follow change. Things were against people coming into Tokyo with families, but I dared."

Seven engineers made the trip with Ibuka, their confidence in him matched by his sense of duty to them. Within a few weeks of Japan's surrender, he rented a small space in the charred recesses of a bombed-out department store on Tokyo's once-and-future shopping boulevard, the Ginza, and set up shop as the Tokyo Tsushin Kenkyujyo (or Tokyo Telecommunications Research Institute). There was no heat and no furniture to speak of, and Ibuka had to send scavenging parties out into the countryside for rice and vegetables. But he had the tools and parts necessary to continue making vacuum-tube voltmeters for the government, and his engineers could find supplementary work as roving radio repairmen at a time when new radios were beyond most people's means although the hunger for information was intense. One way and another, with Ibuka occasionally paying his employees from his savings (about sixteen hundred dollars to start with), Tokyo Tsushin Kenkyujyo made it through the winter, and by spring Ibuka was entertaining the idea of turning his laboratory into a factory.

During the war he had worked on a variety of advanced research projects, including a proposed heat-guided missile, a system that made the Japanese Army's telephone network in China tapproof, and a device that enabled airplanes to detect submarines, magnetically, from up to sixty feet above the sea. ("Twenty-six submarines found," Ibuka recalled. "Destroyed or not, I can't say.") His military assignments allowed him to fulfill an ambition that went back to his student days in the early thirties: to combine the hitherto-separate disciplines of electronics (or "weak currents," as it was known) and mechanics. But Ibuka was frustrated by the need to conform to specifications laid down by bureaucrats. He thought it would be more stimulating to make products for ordinary consumers. The public would be "a stern judge—but fair," he felt. In later years Ibuka was to say that the American electronics industry had been spoiled by its heavy commitment to space and defense, where money and complexity were no object. "So they have no idea how to apply high technology to consumer field," he explained. "But I changed idea. 'First for consumer' is my idea."

He thought he saw partner material in a young ex-navy lieutenant, Akio Morita, whom he had met at the tail end of the war, when both had served on a secret military research team. The heir to a sake-brewing company in Nagoya, Morita had grown up in an unusually wealthy home, filled with such Western appointments as a GE washing machine, a Westinghouse refrigerator, and a Victrola—one of the first in Japan. He was a tinkerer by nature, and when he graduated from Osaka University in March 1944, his physics professor recommended him for the navy research division, which, as they both realized, was a way to survive the war.

That goal accomplished, Morita's next priority was to avoid going into the sake business as the eldest males in his family had been doing since the seventeenth century. He accepted a lectureship in physics at the Tokyo Institute of Technology—a measured act of rebellion which allowed his father to hope for his eventual enlistment in the family brewery. Morita's job turned out to be more temporary than anyone could have expected; he lost it after a few months, when Douglas MacArthur, the virtual dictator of postwar Japan, banned former military officers from public employment.

Morita was secretly pleased. He had already been serving as a kind of part-time consultant to Ibuka, and he found the work a lot more stimulating than his teaching duties. MacArthur's order gave him an easy way out of the teaching job. But there was still his father to reckon with. For a firstborn son to forsake an established family business was a grave step. Ibuka and Morita got on a night train to Nagoya—an extremely cold train, missing some of its windows—and put the case to Morita's father directly. They came away with more than they had asked for; Mr. Morita not only agreed to let his son go to work for Ibuka's enterprise but became one of its main backers. Akio Morita's younger brother Kazuaki eventually took over the family business.

Tokyo Tsushin Kogyo (or Totsuko for short; in English, the Tokyo Telecommunications Engineering Corporation) was established in May 1946. (The name Sony—derived from the Latin *sonus,* meaning "sound," with a nod toward the impish American expression *sonny boy*—came in the fifties.) Ibuka set forth his goals in a prospectus and a declaration of management policies. The company would seek to bring the fruits of the most advanced military and academic research to ordinary Japanese citizens, to aid in the reconstruction of Japan and the refinement of its culture, and to establish an informal workplace where initiative and creativity could blossom. It would avoid growth for growth's sake, "untoward profit-seeking," and all forms of compartmentalization that restricted the exercise of human ingenuity.

> We shall seek a compact size of operation through which the path of technology and business activities can advance in areas that large enterprises, because of their size, cannot enter [Ibuka wrote]. We shall be as selective as possible in our products and will even welcome technological difficulties. We shall focus on highly sophisticated technical products that have great usefulness in society, regardless of the quantity involved. Moreover, we shall avoid the formal demarcation between electricity and mechanics, and shall create our own unique products coordinating the two fields, with a determination that other companies cannot overtake.

In the Japan of the late 1940s, however, such ideals were easier to propound than to fulfill. Totsuko's first consumer product, an electric rice cooker, was a decidedly low-tech affair consisting of a wooden tub and an aluminum heating element. Because of its primitive construction, it produced undercooked rice or gruel as often as it did the job right. A hundred rice cookers were made. Not one was sold. The company had better luck with an electric cushion; but the absence of such amenities as a thermostat made it an obvious fire hazard, and Ibuka refused to put Totsuko's name on it. (The label identified the cushion as the product of the fictitious "Ginza Heating Company.") Totsuko took more pride in a steel phonograph pickup, which not only sold well but gained a reputation for excellence. The necessary steel was obtained by scavenging in the debris of demolished buildings. For basic sustenance, however, Totsuko continued to depend on professional products, and it continued to deal with the bureaucracies that Ibuka yearned to be free of, the biggest of them being the Japan Broadcasting Corporation, NHK, which was run under the close supervision of the United States Army since the Americans regarded radio as a vital element of the democracy that Japan, under their tutelage, was to become.

At the end of 1946 Totsuko moved to a dilapidated warehouse in Shinagawa, a mostly residential neighborhood on the south side of Tokyo. When the company received a contract from NHK to build a radio mixing console, an American major by the name of Whitehouse came to call. He found several dozen men crowded over tables made of boards and boxes in a building whose roof leaked so badly that umbrellas had to be posted over some of the desks when it rained. "Our plant was like in the middle of a trash can," Ibuka recalled. "My friend was scolded: 'Why did you give your order to such a manufacturer?' That got us excited." The company treated its assignment as a challenge; it would be done better and faster than anybody—even an American major who had seen his share of broadcasting studios back home—could reasonably expect. Major Whitehouse was so impressed with the result that he became one of Totsuko's most ardent champions, and before long the company was a principal subcontractor in the construction of a new national radio system.

But Ibuka was still on the lookout for the product that would propel his company into the consumer market, and from time to time he would burst into his little factory with news of some likely gadget which had just come to his attention. One possibility was a wire recorder. Before the war Kenzo Nagai, a professor of engineering at Tohoku University, had developed a machine that used piano wire to record sound. Working along lines parallel to those of scientists in Germany and the United States, Nagai had also discovered the AC-bias recording technique. An audio signal has to be mixed with something called a bias signal before it is recorded magnetically. Nagai had found that a bias signal using alternating current (or AC) rather than direct current (or DC) made for much cleaner recording. The war had blocked Nagai from proceeding with his wire recorder work, but he still held a valid Japanese patent which was available for licensing. Unfortunately wire had serious drawbacks as a recording material: It was expensive, hard to edit or repair, and impractical for recording more than five or ten minutes at a time. The sound quality also left a lot to be desired. At the end of 1948, on one of his frequent visits to NHK's offices, Ibuka saw—and heard—a machine that recorded sound on tape. Fascinated, he continued to borrow the machine, which was U.S. Army property, and took it back to his factory and showed it to his engineers.

By now Ibuka's enthusiasms had taxed the patience of some of his associates—"particularly the man who was in charge of financing at the time," he said. "But once they listened to the sound of the tape recorder we made the decision that we would develop the same thing."

The tape recorder was one of the spoils of the Allies' victory over Germany. The Allgemeine Elektricitaets Gesellschaft—Germany's version of the General Electric Company—began making Magnetophons, as they were called, in the 1930s, and in the war years a few of them were equipped with AC bias for broadcasting purposes. The result was a dramatic improvement in sound quality—so dramatic that a symphony sounded live to the ears of an American soldier in England listening to German radio at two in the morning, although reason told him that the Berlin Philharmonic would be home in bed at that hour. That soldier, John Mullin of San Francisco, and a handful of other technically minded GIs returned to the States after the war with tape, Magnetophon components, and the invaluable knowledge of how to convert a DC-bias machine—a relatively commonplace and distinctly inferior item—into an AC-bias wonder. With their help, a handful of American companies went into the tape recorder business, and by the end of 1948 American-made tape recorders had found their way to Tokyo.

At Totsuko, tape and tape recording became the domain of Nobutoshi Kihara, who had joined the company in 1947 at the age of twenty-one. Ibuka and Kihara had only the most general notion of what recording tape was made of. The one Japanese reference book in which they could find any information on the subject said merely, "In 1936 AEG Co. in Germany invented a tape recorder, using plastic tape which was coated with magnetic material." The book said nothing about what kind of plastic or what kind of magnetic material. The first question, however, was largely academic, for there was essentially no plastic to be had in Japan at the time, and a national austerity program prohibited imports of anything but fuel and food. The country was a desert island from a supply standpoint. To be an inventor in such a setting, one had to cultivate the skills of a Robinson Crusoe as well as a Thomas Edison. For the tape, the only material that was both available and remotely appropriate was paper, so paper it was, although by rights recording tape should be thinner, smoother, and stronger than paper is disposed to be. For the magnetic coating, Kihara eventually chose an iron oxide powder obtained from a drug wholesaler of Morita's acquaintance, mixed it with lacquer, and, with a spray gun, sprayed it onto his paper tape until he had a medium that could be wound through a jerry-rigged wire recorder to produce a thin approximation of human speech.

The need to make tape out of paper, and to construct a recording machine gentle enough not to break it, contributed to a growing mania for exactness and reliability at Totsuko. A major problem in adapting a wire recorder for tape is the need to maintain constant speed as the tape passes the recording head, while the diameter of the tape on the pickup reel is steadily increasing. Kihara installed a rubber roller near the head to move the tape through at a steady speed; later he saw the same device on an American model. He was inventive, methodical, and good with his hands—a born problem solver—and he enjoyed the feeling that no solution was too outrageous for Ibuka. "At an early stage of the development of the tape recorder, we had a problem with the motor," Kihara recalled. At the time we had to use vacuum tubes, and in order not to affect the vacuum tube, we had to use a very quiet motor. That was so difficult that we thought about using a spring-powered motor instead of electricity. When I proposed that idea to Mr. Ibuka, he said, 'Go ahead—do it!' An extravagant idea such as that, he picked up on." Kihara and his fellow engineers went to great lengths to please the boss. "I prepared many new things in my laboratory and didn't show them to Mr. Ibuka," Kihara said, "so that when he came to me and said, 'How about this or that?' we already had some sort of prototype, and if we change it a little and show it to him next day, he thinks it was done overnight, and he is astonished. I like to see the face of Mr. Ibuka astonished."

By October 1949 the tape recorder work was going so well that the company decided to spend twenty-five hundred dollars—a huge sum by its reckoning—on the rights to the AC-bias patent. In January 1950 Totsuko put its paper-tape machine, the G-Model Tapecorder—Japan's first tape recorder—on the market. Ibuka and Morita thought they had come up with an irresistible product, and, in a sense, they had. When Morita explained what a tape recorder could do, prospective customers were invariably eager to hear for themselves. Sometimes Morita would be invited to a banquet to give a demonstration, and once there he would be asked again and again to record and play back ballads sung by geishas and their patrons. "Everyone said it was wonderful to hear his own voice," Morita recalled, "but no one wanted to buy it." He began to feel like a *taikomochi*—a jester.

In an expensive curio store one day Morita looked on in amazement as a customer—exclaiming, "What a bargain!"—walked off with a carved ivory figurine which cost even more than a G-Model Tapecorder. That scene helped channel Morita's thoughts. Value, he realized, was a matter of perception. People had to understand the uses of a tape recorder for it to become a thing of value, and Totsuko had to understand those uses itself before it could communicate them to the public. More than six months passed before the first substantial order. It came from the Japanese Supreme Court, which saw the tape recorder as a way to compensate for a shortage of trained stenographers. Other early buyers included magazines, newspapers, music schools, and the national police. An American pamphlet entitled *999 Uses of the Tape Recorder* made a fortuitous appearance during this period. Morita and his colleagues were surprised to learn that most of the recommended uses were in teaching and research. The pamphlet identified astronomers and microbiologists as promising customers because, it explained, they needed a way to note their observations without taking their hands off their telescopes and microscopes.

Audiovisual instruction, it so happened, was a practice the occupation forces were trying to inculcate in Japanese schools. The Americans, perhaps doubting that Japanese teachers could be trusted to impart the new value structure without outside prompting, had imported several thousand sixteen-millimeter movie projectors, and they were encouraging the use of piped-in radio broadcasts in classrooms. A tape recorder ought to have a place in such a scheme of things, Morita and his associates reasoned. But the G-Model weighed a hundred pounds and cost 160,000 yen—about twice the typical annual salary in Japan at the time. Even for a school, it was too much to handle.

That summer Ibuka took his best engineers to a hot springs resort near Yokohama for a ten-day brainstorming session. By December the H-Model—one-fourth as big, half as expensive, and amiably packaged in a striped suitcase—joined an improved G-Model on the assembly line, and the employees held a spontaneous celebration.

As one of the few people in the company with a driver's license, Morita took to the road to promote the H-Model's attractions. He and a colleague drove around in a Datsun truck and held forth to audiences of teachers, explaining how the tape recorder worked and how it could be used. They came away with lessons to be applied in future sales sessions (and in refinements of the product) and with a growing number of customers. Eventually most of Japan's forty thousand schools saw fit to buy a Tapecorder, and as Kihara struggled to improve the tape and the sound quality, it gradually became standard equipment at the local radio stations that Totsuko was helping get started. And the radio stations, no longer so reliant on live broadcasts, proliferated.

The tape recorder was the golden goose of Totsuko's operations through most of the 1950s, and the company guarded it jealously, hollering bloody murder when a few American machines were sold to Japanese customers, taking one Japanese competitor to court when it claimed to have discovered a "new" AC-bias technique, and, in 1955, securing a five-year extension of the AC-bias patent (to compensate for the lost years of the war) just when a host of Japanese companies were gearing up to enter the field.

In 1956, when battery-run motors were still in their infancy, the company came out with a battery-operated tape recorder, the Babycorder. It was a portable machine, shaped like a thermos, which economized on size and weight by requiring the user to do his rewinding and fast-forwarding by hand. A year later Bunichi Hirose, the president of a Japanese confectionery company, attended a Boy Scout jamboree in Birmingham, Alabama, toting his Babycorder. The prime minister of Great Britain, Harold Macmillan, was on hand, and when he saw the Babycorder, he asked what it was and where it hailed from. To the second question Hirose gave an answer that Macmillan found unacceptable. "They can't make something like that in Japan," he declared. Hirose recorded Macmillan's speech and played it back for him, and—according to the journalist Yasuzo Nakagawa, in his book *The History of Magnetic Recording in Japan*—"Macmillan opened his eyes wide in surprise and liberally praised the excellence of Japanese technology."

Masaru Ibuka made his first trip to the United States—it was also his first trip outside Japan—in March 1952. In New York he learned almost by chance of AT&T's decision to license the transistor, a discovery that had emerged five years earlier from Bell Laboratories, the AT&T research unit. Ibuka had come to America with his mind on selling tape recorders, and he did not immediately see the significance of the transistor offer, which called for a payment of twenty-five thousand dollars up front. But one night he was up late, clearheaded and unable to sleep, when the transistor appeared to him in a vision. Totsuko was selling most of its tape recorders to schools and government offices, which tended to make their purchases on a seasonal basis. Ibuka was still on the lookout for a true consumer product and for something that would sell year-round. He was also worried about the young engineers he had hired to work on tape recorder development. He needed a new goal to keep them busy. As a smaller, sturdier, and less power-hungry substitute for the ubiquitous vacuum tube, the transistor promised to make it possible to build sophisticated electronic devices for casual, everyday use. It was the embodiment of the goal that had led Ibuka to found his company in the first place: making high technology available to the general consumer.

In the large and prosperous community of people who make their livings these days by pontificating on the decline of American manufacturing, there are two broad factions: the America Bashers and the Japan Bashers. The former spend much of their time deploring the failure of American industry to follow up on American discoveries, and they tend to make it sound like a case of wanton neglect. But there are also differences of outlook and taste that help explain why one society takes to a certain technology more readily than another. Compactness—a major element of the transistor's potential—was a quality that held little appeal in the United States of the mid-fifties. The Japanese, by contrast, were a people who had lived for centuries with a scarcity of resources and habitable land and had made the most of it. Over the years they had honed their talents for detail work on bonsai trees, boxes within boxes, finger-size dolls, and other magnificent feats of minusculity. The empress Jingu, supposedly inspired by the wings of a bat, is said to have invented the folding fan in the third century A.D.

The flat fan had come over from China and Korea, and Japan sent it back collapsible. In the fourteenth and fifteenth centuries the Chinese exported Japanese folding fans to Europe, thus making them the "first Japanese good to dominate the world market," as a Korean writer, O-Young Lee, has observed. Other symptoms abound of the national obsession with smallness: the adoption of the cherry blossom, slender in life-span as well as size, as a symbol of beauty; the haiku, "the world's shortest poetic form," according to Lee; and the *bento,* the box lunch that verges on a work of art. For such a nation the transistor was the perfect toy.

Ibuka tried and failed to make an appointment at Western Electric, the AT&T subsidiary that had custody of the transistor rights. But he left instructions with a Japanese friend in New York City to follow up on his inquiries. On his return to Japan, Ibuka learned from his friend that the twenty-five thousand dollars were an advance on royalties rather than an outright payment, as he had supposed. His enthusiasm mounted.

Japan Bashers, like American Bashers, are fond of sermonizing about simple virtues or the lack thereof, and in their accounts of Japan's postwar economic successes much is said about the government's Svengali-like role as a winnower, funder, coordinator, and promoter of promising companies and industries. The case of Sony and the transistor makes for a problematic chapter in treatises on this theme. Despite the company's willingness to pony up a twenty-five-thousand-dollar advance, it had to have approval from MITI, the Ministry of International Trade and Industry, to spend so much money overseas—and MITI, an agency created after the war to stimulate capital-intensive and technology-intensive industries, was not in an approving mood. For two years the application remained in limbo because of the opposition of a minor MITI power broker who, far from seeing an opportunity to build a new Japanese industry, saw a sorry combination of a fledgling company and an unproved technology. Toshiba, Mitsubishi, and Hitachi—all of them much bigger and more established companies than Totsuko—were preparing to enter the transistor business with technical guidance from RCA. If companies of that caliber needed American help, the MITI official said to himself, what business did Totsuko have thinking it could do the same thing unassisted?

Forced to defend the worthiness of his company, Ibuka pointed out that Western Electric had investigated Totsuko's financial and technical wherewithal and seemed to feel that it measured up. To prove his point, he showed MITI the tentative deal (expressly subject to MITI's approval) that Western Electric and Totsuko had negotiated. His ploy backfired. Instead of being impressed, MITI officials chewed Ibuka out for agreeing to *any* contract, even a tentative one, without their say-so. Totsuko's behavior was "inexcusably outrageous," Ibuka was told.

Morita made his first trip to the United States in August 1953. "Why did Japan go to war against such a big country?" was his first reaction. He had come with thoughts of exports on his mind, but the high quality of American manufactured goods was demoralizing. How could a small company in a primitive country like Japan make anything that Americans would want to buy? He flew on to Europe, and a stop in Germany—a nation that seemed unaccountably prosperous in defeat—fed his doubts. "I was walking in Düsseldorf," Morita recalled, "and it was so hot that I sat down in a café and asked for an ice cream. The ice cream came with a small bamboo umbrella inside, and the waiter said, '*This* is from your country!' I was really shocked that such a thing could represent Japanese products." Holland revived Morita's confidence. The sight of farms and people on bicycles reminded him of Japan, and in their midst, in Eindhoven, he visited the Philips Corporation, which had grown over several decades from one man's new venture into a huge and internationally renowned maker of high technology. "What has been done by Dr. Philips can be done by us," Morita decided.

In their conversations with the people at Western Electric, Morita and Ibuka had been vague about the reasons for their interest in making transistors. A Western Electric engineer had advised Morita that a hearing aid might be an appropriate transistorized product for Totsuko to attempt. "Well, yes," Morita had answered. But in truth he couldn't see much of a market for hearing aids. Nor could Ibuka, who thought he had a better idea.

"Radios," he announced. "We're going to use this transistor to make radios—small enough so each individual will be able to carry them around for his own use, with power that will enable civilization to reach even those areas that have no electric power yet."

In January 1954 MITI finally relented, and Ibuka was off to the United States to seal the deal with Western Electric and learn about his newly acquired discovery. Only now did Ibuka dare to mention the word *radio,* which provoked a paternal lecture from the Western Electric people. The transistor was a fascinating discovery, they told him, but its near-term applications were limited. So far no one had been able to manufacture a transistor that could come even close to handling the high frequencies a radio would require. A hearing aid was the thing, Ibuka was advised. The only transistorized consumer product actually on the market at the time was a Zenith hearing aid. It cost nearly five hundred dollars, and the austerity measures of the Japanese government limited overseas travelers to a budget of fifteen dollars a day, so Ibuka could not even afford to *buy* a hearing aid.

Kazuo Iwama, who became Totsuko's resident transistor authority, remained in the States for months, sending voluminous reports back home on Western Electric's production techniques, while his fellow engineers in Tokyo took up the work of trying to make a transistor that could handle the higher frequencies and be manufactured at a manageable cost. It was an exceedingly expensive and, as it seemed at the time, slow business—one that would not have been possible without the profits from Totsuko's monopoly position in the Japanese tape recorder world. But the need to make major gains in a component's yield, or output, was not unfamiliar to the company's engineers. They had been through similar trials with the paper-based recording tape of their early tape recorders. *Gaman,* or persistence, is a highly valued quality in Japanese life. "To challenge the yield is a very interesting point for us," Ibuka said. "At that time no one recognized importance of it."

By early 1955, having improved the high-frequency yield of its transistors in small but steady increments, the company was ready to market a radio—a squat little thing, slightly smaller than a toaster, with the word *Transistorized* emblazoned across the top, and, in less conspicuous Art Deco-style letters above the tuning dial, the name Sony, which soon became the company name. It was not quite the world's first transistor radio. That honor fell instead to an American company,

Regency, using transistors made by Texas Instruments. But Sony was only a few months behind, and had it not been for the shortsightedness of the Japanese government the outcome might well have been reversed.

Morita made another trip to the United States in the spring of 1955. His first visit had been exploratory. This time he came to sell, and he brought an assortment of microphones, tape recorders, and radios. But he was not prepared to take just anybody's money. The cameras made by Canon and Nikon were the only Japanese products with a reputation for quality at the time. Morita was determined to add Sony's products to that select list. He wanted to see them in department stores, music shops, and other suitably respectable outlets instead of the discount stores where Japanese goods could generally be found. Bulova, the watch company, offered to buy a hundred thousand transistor radios—an astonishing number which, in money terms, came to several times Sony's entire capitalization. But there was a condition. The radios would have to be sold with the Bulova name on them. "Nobody in this country's ever heard of Sony," Morita was told. He decided to turn the Bulova order down. "On no account should we permit the use of their brand name," he told his colleagues in Tokyo when he called to discuss the offer. "We've got our own, Sony, and we should stand by it."

With the uneasy backing of the home office, Morita reaffirmed his position in a meeting with the president of Bulova, who (according to Morita's recollection) responded scornfully. "Whoever's heard of Sony?" he said. "Our brand has a world reputation with fifty years of history behind it."

"Fifty years ago how many people knew your name?" Morita replied.

Two years later Sony announced the introduction of the world's first shirt pocket-size radio: the TR-63. Actually it was a little larger than that, but Morita had the company make a foray into the shirt business to provide its salespeople with radio-size pockets, and the world, if not entirely fooled, was entranced. In the years that followed, Sony sold a million of its "pocketables," along with hundreds of thousands of transistorized models that came with AM/FM and other refinements. For two years Sony had the pocket-size field to itself, worldwide. Even when imitation radios started appearing in the United States with brand names like Sonny and Somi and prices of $12 and $13 instead of Sony's $39.95, its models remained best-sellers, helped along by a reputation as "the radio that works."

The tape recorder and the transistor radio were formative experiences for Sony. They established a pattern: the determination to do something other companies had not done; the struggle to improve the yield; the maniacal embrace of a goal ("If Sony fails in any one thing, it is damned," the journalist Yasuzo Nakagawa has said); and the idea of "educating the market" to the virtues of a product born of intuition rather than market research.

Over the years Morita would often be asked, "Where is your market research?" He would point to his forehead and proclaim, "Here!" The "microtelevision," for example, was conceived in the knowledge that RCA had tried and failed to sell a seven-inch TV set; indeed, it had become an article of faith with American TV makers that smallness was the wrong way to go. The early sixties were the age of the console TV, the "home entertainment center," and Earl ("Mad Man") Muntz's twenty-seven-inch monster. Into that setting Sony launched nine-inch, five-inch, and four-inch sets which ran on batteries as well as house current, and Doyle Dane Bernbach—whose Volkswagen ads had impressed Morita as unusual and amusing—ebulliently promoted their uses. Now you could watch TV in the kitchen, or on a fishing boat, or in bed with the set resting on your stomach. "Tummy Television" the ads called it.

When Tummy Television came to America, so did Morita. To apply Sony's marketing ideas properly in this critical market, he decided, he should be on the scene personally, overseeing the creation of an American sales operation which (unlike those of most foreign manufacturers in those days) would be a direct subsidiary of the parent company. Few out-of-towners have every landed in New York City more splendidly. To rub shoulders with American executives it was necessary to live like one, Morita concluded. So in June 1963 he moved his family into a twelve-room apartment on Fifth Avenue opposite the Metropolitan Museum, promptly sent his two sons to summer camp in Maine for a crash course in the English language and American ways, joined a country club—golf being an activity for which, like many Japanese executives, he needed no preparation—and began attending operas, concerts, and plays, and hosting dinner parties on a grand scale. On the theory that people had to see a new product work before they could really appreciate it, he opened a Sony showroom, also with a Fifth Avenue address.

Ibuka and Iwama were skeptical of the need for such a costly and dramatic transplantation, and Morita's expansive life-style and growing fondness, in the ensuing years, for German cars, Italian loafers, and French helicopters probably caused a number of unwesternized eyebrows to arch. But the results were hard for his colleagues to criticize.

And there was very little criticism from the American business world. By the late sixties Sony had sold a million transistorized TVs to American consumers. It was being celebrated as a miracle company—and Morita as its miracle man.

> Since its origins in 1946, backed by only $500 in capital, Sony has never really had a bad year [*the Wall Street Journal* observed in November 1969]. . . . Although Sony is far from the largest Japanese electronics concern, it is No. 1 with American investors; they own nearly 29% of the outstanding stock [and] it's a very pleasant holding at the moment. . . . All in all, Sony has an enviable reputation for growth and technological innovation, for shrewd marketing sense and for willingness to tackle the tricky job of selling unfamiliar products to the American consumer on his own home ground. Less apparent to Westerners and more significant for world trade, however, are the radical internal changes that Sony has made in managing its operations in Japan, under the influence of the cosmopolitan Mr. Morita.

It was the *Americanness* of Sony that impressed the *Journal* the most. The company was praised for such un-Japanese practices as bypassing school records in its personnel decisions, welcoming eccentrics, hiring talent away from competitors, and, above all, moving fast. In December 1965 Sony agreed to help IBM get into the business of making magnetic tape for computers. Sony built a plant in Japan, flew it in pieces to Boulder, Colorado, assembled it, and put it in operation by the fall of 1967—about "half the time we normally would allow," an

IBM executive said. Harvey Schein was the president of the international division of CBS at the time, and he spent three years trying to persuade another Japanese company to set up a joint venture with CBS to sell records in Japan. Fed up with the delays, he finally arranged an exploratory lunch with Morita, and "by the time we'd had soup, we had a deal," Schein liked to say.

American law required Sony products to carry the legend "Made in Japan," but Morita made sure the words were no bigger than necessary. (On one occasion, customs agents ruled that they were somewhat smaller than necessary, and rejected a shipment as a result.) It was no coincidence, either, that the Sony name itself had a certain ethnic ambiguity about it. To an American ear, it sounded no more foreign than Xerox or Kodak. Indeed, the name seemed so American to most Americans that, according to a survey by *Fortune* magazine, a high percentage of retailers who carried Sony products claimed they had never sold anything made in Japan, and never would.

GENERAL SARNOFF'S BIRTHDAY PRESENT

ON SEPTEMBER 27, 1951, David Sarnoff celebrated the forty-fifth anniversary of his start in the radio business, and when Sarnoff celebrated something, so did the Radio Corporation of America, which he headed. The ceremonies, held at the RCA Laboratories in Princeton, New Jersey, consisted of a lunch featuring speeches and congratulatory telegrams (including the obligatory missive from President Harry S Truman), followed by the rededication of the Princeton facility as the David Sarnoff Research Center, the unveiling of a bronze plaque bearing Sarnoff's likeness, and the distribution of commemorative medallions. Then General Sarnoff—he had received the rank in 1944 for services as Dwight Eisenhower's communications consultant—gave a talk to the attendant engineers on a formidable question: how to prepare for the fiftieth anniversary of his start in the radio business, just five short years away.

He suggested three presents that he would be pleased to receive on that occasion: "a true amplifier of light," "an all-electronic air conditioner," and "a television picture recorder that would record the video signals of television on an inexpensive tape. . . ." Sarnoff had been celebrated for his powers of prognostication as far back as 1916, when he had envisioned the presence of "radio music boxes" in ordinary homes. He delighted in constructing lists of technological wonders that somebody or other (preferably in the employ of RCA) was bound to invent by such and such a time, and if the lists had taken on a grab-bag quality in recent years, it was partly because Sarnoff had found that no one remembered the predictions that didn't come true.

Sarnoff's first two requests caused some bewilderment among the engineers, but they understood the third one well enough. RCA, through its subsidiary NBC, was in the business of running a TV network, and the only way to store a TV program in those days—the days of live TV—was by kinescope recording, a process that amounted to putting a movie camera in front of a TV screen. Kinescope recordings were expensive and not very pretty to look at. They also took time to make. It was no easy trick to take the live transmission of a program that went on the air at, say, 8:00 P.M. in the East and be ready to broadcast the kinescope version three hours later in California—in other words, at 8:00 P.M. Pacific time. A videotape recorder, if such a thing could be built, promised to be a simpler, cheaper, and quicker answer.

In principle, there was no barrier to recording a TV signal on magnetic tape. The difference between the demands of audio recording and video recording was essentially one of degree. A TV image can be as complicated as a checkerboard with, say, two hundred thousand squares. Because the image changes thirty times a second to simulate motion, the TV signal has to convey millions of instructions to the picture tube every second. Thus, in American television the video signal covers a frequency range of roughly four million cycles a second, while in audio twenty thousand cycles a second will do the job—even less, if voice rather than music is what's being communicated.

Recording tape consists of a base (made of plastic or, in the case of Sony's early tape, paper), a metallic powder, and a coating, or binder, that holds the powder to the base. In the standard audiotape recorder the tape is pulled past a recording head, a pincer-shaped electromagnet, which produces fluctuations of intensity and polarity in response to the electrical input from a microphone (or some other outside source), leaving a longitudinal trail of magnetism on the tape. On playback the roles are reversed. The magnetized particles on the tape dictate to the head, which relays the fluctuations in their magnetism to an amplifier and from there to a speaker. The problem confronting the pioneers of video recording was to figure out how to adapt this relatively straightforward technology to a much more intricate signal, and the solution that came to most of them was simply to speed the process up. Just as a tape recorder is set to a higher speed for music than for speech, it would have to be set to a still higher speed—a much higher speed—for TV.

The first prototype of a videotape recorder—demonstrated at Bing Crosby's recording studios in Los Angeles on November 11, 1951—was a modified audiotape recorder in which the tape ran at a hundred inches a second and a single reel held sixteen minutes of programming. The picture suffered from poor resolution, flicker, diagonal disturbance, and occasional ghosts, but Crosby was sufficiently impressed—and sufficiently aware of the economic implications—to commit himself to the goal of developing a broadcast-quality videotape recorder and to sink his money into the project for the next four years.

Almost all the engineers who worked on the problem in those years—and there were plenty of them, in Britain and Japan as well as the United States—took a brute-force approach. If a slightly souped-up audio recorder wouldn't do the job, they would soup it up some more. At RCA the quest became the in-house equivalent of the Manhattan project, and the prototypical tape was soon hurtling past the prototypical head at a speed of thirty feet per second, or twenty miles an hour—so fast that during a public demonstration in the fall of 1953 the engineer charged with stopping and rewinding the machine had to wear heavy leather gloves (which served, in effect, as brake drums). It took well over a mile of tape on that year's model to hold a four-minute-long recording. The arithmetic was impossible, and the picture wasn't much better. On the eve of the demonstration Sarnoff complained of a fuzziness in the image, and his engineers, for want of any other remedy, were up late that night removing the front row of seats to keep the press and Sarnoff at a visually safe distance from the monitors. The next morning, according to George Brown, an RCA engineer, Sarnoff "asked how this obvious improvement had been accomplished." He was told that the "boys" had worked on the problem until midnight.

While the tape speed in the RCA design was inching up on Mach 1 and the publicity was keeping pace, a few young engineers at the Ampex Corporation in Redwood City, California, were trying another approach. Ampex, founded in 1944 by a Russian émigré, Alexander M. Poniatoff (who added *ex* for "*excellence*" to his initials to get the company name), was a tiny blip on the American corporate horizon—a collection of talented engineers who had built the United States's first professional audiotape recorder, permitting radio stars like Crosby to be on the air and playing golf at the same time. (In 1944 Crosby quit the "Kraft Music Hall" because of an inviolate network policy—the network was NBC—against recorded shows. A year later he returned to the air on ABC, a newly established network which, in its hunger for talent, agreed to let him use acetate discs. But the needle scratches and poor frequency response of the discs caused complaints, especially when third- and fourth-generation copies were used in order to facilitate the removal of errors or weak passages—or off-color jokes, a Crosby specialty—in the original performance; and the ratings suffered. When Ampex came out with its tape recorder, in April 1948, Crosby was the first customer.)

The engineer in charge of Ampex's videotape recorder project, Charles Ginsburg, was committed to a concept originally suggested by Marvin Camras, the holder of the American patent on the AC-bias technique. What Camras had envisioned, and Ginsburg was trying to build, was a machine in which the head as well as the tape was a moving part—in fact, the faster-moving of the two. Camras had sketched out a recording system with three heads mounted on the flat side of a rotating drum, laying short, semicircular tracks across the width of the tape instead of a continuous, straight-line tract along the length, as in audio recording. Tape is by nature an inconstant material, prone to slipping and stretching, and it was difficult to keep the tape under control at the high speeds contemplated by the Crosby and RCA engineers. The Ampex approach would make it possible to slow the tape speed down to a more manageable rate, while still achieving an extremely high "writing speed"—the relative speed with which the tape and head went past each other. This, in turn, would make it possible to accommodate a much longer recording on a single reel.

But the virtues of rotating heads remained largely abstract through two years of on-again, off-again development. Because of the discontinous track, the responsibility for transmitting the signal had to shift from one head to another a few hundred times a second—another new idea, and one that created all kinds of problems. A nineteen-year-old prodigy, Ray Dolby (later of "Dolby sound" fame), worked on the project from August 1952 until he was drafted into the army the following spring, and he came up with the notion of using four heads instead of three, along with a two-way switching system that simplified the circuitry and made the timing more exact. Later the heads were moved from the flat side of the drum to the curved side, turning the tracks into straight lines instead of arcs—an arrangement that became known as transverse scanning. But the signal-switching process continued to make itself felt in the picture. "The odious label 'venetian blinds' came into being," Ginsburg recalled, "to describe certain very unpleasant flaws at the points representing the crossover from one head to the next. . . ."

Like their fellow engineers at RCA, Bing Crosby Enterprises, and the British Broadcasting Corporation, Ginsburg and his associates were using a form of amplitude modulation, or AM, to record the video signal on the tape. As any radio listener knows, AM is more vulnerable to interference than FM, or frequency modulation. FM, on the other hand, had never been used in magnetic recording because it took up a great deal of space, and the extremely high-frequency range of a video signal would only aggravate that problem. Using AM, RCA had arrived at a video recorder with a tape speed of twenty miles an hour and a recording capacity of four minutes a reel. If those numbers were going to get better instead of worse, FM was out of the question—or seemed so until December 1954, when a colleague of Ginsburg's, Charles Anderson, had a far-out idea. Before coming to Ampex, Anderson had worked in the comparatively new field of FM broadcasting, so he understood the subtleties of FM better than anyone else on the VTR development team. Just because the video signal had to cover a frequency range of four million cycles a second, it didn't necessarily follow, Anderson argued, that the signal used to record it on tape had to occupy such a broad band width. Maybe an FM carrier signal set to a much lower frequency and modulated within a much narrower range would suffice. After all, he reasoned, it was the *rate of change* in the frequency, not the frequency itself, which conveyed the picture information. Anderson's idea was hard to comprehend, but a similar approach had been successfully adopted in AM radio transmission, and the Federal Communications Commission had studied the technique for use in TV broadcasting. Within four weeks Anderson had worked his proposal into a prototype that delivered a promising picture. Transverse scanning had been a lonely idea, waiting for another idea to rescue it from a pointless existence. Together, transverse scanning and FM recording made it possible for Ampex to produce a broadcast-quality videotape recorder in which the tape moved at the relatively stately pace of fifteen inches a second and a ninety-minute program could be recorded on a fourteen-and-a-half-inch reel (as opposed to four minutes on a seventeen-inch reel in the RCA prototype).

After many further refinements the management of Ampex was invited to a demonstration in February 1956. "We recorded for about two minutes, rewound and stopped the tape, and pushed the playback button," Ginsburg recalled. "The group, which had been quiet up to that point, suddenly leaped to their feet and started shouting and handclapping."

When a number of companies are working on the same problem, progress is usually a subtle and incremental affair in which no one ever gets very far in front of the pack. But when Ampex gave the first public exhibition of its machine in April 1956, at the National Association of Radio and Television Broadcasters convention in Chicago, the finality of its achievement was immediately apparent to everyone in the industry, and every other engineer who had labored on the problem knew that his labors had been essentially for naught.

The people at Ampex, however, had only the dimmest notion of what they had done. For planning purposes, it behooved them to estimate the potential demand for their new product, and they concluded that they could expect to sell thirty machines in four years. After all, as Joseph Roizen, an Ampex development engineer and product manager, explained, "It was just going to be used for time delay, and how many people needed to do that?" No one, apparently, anticipated that videotape recorders—or VTRs, as they came to be called—would end the age of live TV and that in an age of taped TV every station and production company in the land (and other lands) would need them in quantity. Within fours days of the Chicago demonstration Ampex had contracted to sell some eighty machines at fifty thousand dollars each.

At RCA Ampex's triumph was greeted with the same good sportsmanship that the United States government had shown toward the Soviet H-bomb. RCA's executives had to restrain their fury in the presence of their leader, however. No one wanted to tell David Sarnoff that a tiny company in a place called Redwood City—an outfit that NBC people liked to refer to as "Hobby Lobby"—had run off with his birthday present. And so, for a time, no one did. When the fiftieth anniversary of Sarnoff's start in the radio business rolled around that September, the general's aides presented him with another RCA-made, fixed-head VTR, and they made no effort to disabuse him of his faith that the RCA machine would yet emerge as a world standard. Another few months passed before they could bring themselves to say that the VTR project was not going as well as they had hoped and that, on the whole, RCA might be wise to scrap its own design and work out an arrangement with Ampex instead.

EAST IS EAST

FOR SEVERAL YEARS Ampex basked in the afterglow of its triumph. Some discoveries hold up better than others, and Ampex's held up extremely well from both a legal and a technical standpoint. What Ginsburg and his colleagues had fashioned, it turned out, was not just *one* method of recording a TV signal but, in terms of picture quality, the best method within reach. The key elements of their system—transverse scanning and FM recording—lent themselves to strong, clearly defined patents, which would allow Ampex to control the VTR business for years to come. Or so the company's lawyers assumed.

In Japan, however, a movement for technological self-sufficiency was gaining ground at the expense of some of the principles on which Ampex's lawyers based their calculations. In 1958, when NHK and other Japanese broadcasting interests applied to the government for permission to buy Ampex VTRs at a cost of more than eighty thousand dollars each, the thought of so much money flowing overseas provoked MITI to make a pool of money available for a crash effort to produce a domestic alternative. At MITI's suggestion, a VTR roundtable was created to coordinate the research, and a number of companies, including Sony, participated.

General MacArthur and his minions had by no means neglected the concept of antitrust in their lesson plan for the new Japan, but it had failed to take root despite their best efforts. As a young man Morozumi Yoshiko, who later served as a vice-minister of MITI, spent a difficult part of the year 1947 translating an anti-monopoly law, article by article, from the English in which it had been written into good statutory Japanese. "We didn't really know what they were talking about," he said later.

At Sony Nobutoshi Kihara had been tinkering with VTRs in a casual way since 1953. He and his fellow engineers responded to MITI's challenge by building a perfectly acceptable VTR in a period of four months. But the Sony VTR was a virtually exact copy of Ampex's, and Ampex, which had filed for Japanese patent protection for its VTR design, was not about to invite a local competitor into the business. It wanted the Japanese broadcast market for itself.

In 1959 another Japanese company, Toshiba, demonstrated a prototype VTR that employed a new recording configuration called helical scanning. In a helical-scanning machine the tape is wrapped in a spiral around the drum, and the recording head lays down a track that runs diagonally across the tape at an angle nearer to lengthwise than widthwise. With helical scanning, each swipe across the tape became considerably longer than in a transverse scanning machine, and it could handle half, rather than a thirty-second, of a frame. This greatly reduced the number of times the signal had to switch over from one head to another, allowing for simpler circuitry and a VTR with only one or two heads instead of Ampex's four. Sony, Matsushita, and the Victor Company of Japan, or JVC, also built helical-scanning VTRs that year, perhaps in the hope that by deviating from Ampex's design they would find a way to escape its patents. But the picture quality on their machines was poor. The elasticity of the tape and the difficulty of maintaining constant tape speed—problems that Ampex had brought under control with transverse scanning—became problems once again. The early helical-scanning VTRs might be acceptable for some purposes, but not for broadcasting, the most urgent purpose. All these machines depended, in any case, on technology developed by Ampex. The FM recording method, and the so-called Ginsburg patent which protected it, still stood as an apparently insurmountable barrier to any Japanese entry into the VTR business.

Ampex, too, was working on a helical-scanning machine, as part of a broad effort to reduce the size and cost of its VTRs. Progress had been slow, however, and in the decision-making ranks of the company there was concern about Ampex's future. "At that time," Joseph Roizen of Ampex recalled, "we had a vice-president by the name of Phil Gundy, who was kind of a showman but a good marketing guy—he knew what the world needed. And what he was beginning to realize in the late fifties was that Ampex had better transistorize its recorders to make them smaller and more reliable and less expensive. This was an era when the engineers like myself who had grown up with vacuum tubes—remember them? those little glass envelopes with the filament that lights up?—were deathly afraid of transistors. We knew how to design circuits with vacuum tubes backwards. We didn't understand solid-state devices. Talk about a mid-life crisis! Most of the guys were, say, in their late twenties or thirties, and all of a sudden everything they had learned in college was obsolete, and young guys were coming out of Stanford and Cal and so on with this transistor technology. So there wasn't anybody at Ampex who was very enthusiastic. In fact, some of the engineers had little signs on their desks that said, 'Help stamp out transistors!' Gundy recognized the problem, and when he was in Japan and saw some of the first Sony transistorized devices, he realized that here was a company that could design the transistor circuits that would replace the tube circuits in Ampex's machine."

With the backing of George Long, Ampex's president, Gundy made a proposal to Akio Morita and Masaru Ibuka, and in July 1960—with typical Sony dispatch—they signed a one-page letter of agreement. Sony would design and supply transistorized circuits for use in a "portable" version of the standard Ampex VTR. In return, Sony would get the right to make VTRs for nonbroadcast customers. It was an unprecedented step in Japanese-American corporate relations, and one taken, by all accounts, with high hopes on both sides. Masahiko Morizono, later a deputy president of Sony, spent months in Redwood City studying Ampex's VTR technology, and a team of Ampex engineers was dispatched to Japan to work with a Sony team on the design of the circuits.

It was not an easy collaboration. Ampex was a company that made complicated and high-quality products for professional users, and its engineers tended to regard their VTR standards as sacrosanct, while the Sony engineers did not enjoy being relegated to the role of subcontractors on a project whose basic outlines were preordained. What's more, they regarded some of the Americans as overly individualistic and not sufficiently conscious of the big picture. In *The History of Magnetic Recording in Japan,* Yasuzo Nakagawa quotes a Sony engineer reading the mind of an Ampex engineer. "I am a servo specialist," says the American. "Therefore, if I make a servo to the given specifications, that's fine." (A servo mechanism is a kind of thermostat for speed.) In the Japanese engineer's scenario, the American engineer discovers he can make a servo that meets higher specifications than prescribed, but instead of pondering how it might affect the overall performance of the product and calling the possibility to his colleagues' attention, he dismisses it as beyond his mandate and keeps the possibility to himself. When he finishes his servo, and it is duly installed in a prototype that produces an unsatisfactory picture, he announces, "It's not my fault. Something else is wrong."

The Ampex engineers, for their part, were more than a little bewildered by the whole experience. They were in Japan on orders from above, and their stays there typically lasted only two or three weeks—hardly time enough to adjust to such a strange setting. As one of them said later, "None of us spoke any Japanese, except for a few words we had picked up. And of all the Sony engineers, only Mr. Morizono was really fluent in English, so most of our communications had to be channeled through him." It was not easy to be on the cutting edge of technology and the fuzzy edge of comprehension at the same time.

But the engineers got along famously compared to their superiors. In 1961 Ampex underwent a change of management. George Long resigned as president after a bad year, and in came William Roberts, an ambitious man who "looked upon management as someone else might look upon the priesthood," according to an acquaintance. Roberts had been recruited from Bell & Howell, and he made no secret of the fact that he had left there because his hopes for promotion had been foiled by the naming as chief executive of another young and dynamic type, Charles Percy, later a senator from Illinois.

Ampex's expectations of Roberts were fairly modest: that he would restore the company's sagging profits. Roberts had considerably higher expectations of Ampex. Its products enjoyed an excellent reputation in their niche of the electronics world—the broadcasting industry, basically—but he was not the kind of man who could be at home in a niche. Bell & Howell, his old employer, made consumer as well as professional products. Except for a few extremely fancy audio recorders (and a line of prerecorded tapes) aimed at the highest reaches of the consumer market, Ampex had scarcely begun to test those waters, and that struck Roberts as a deplorable oversight. With the quality of Ampex's technology, he believed, the superiority of the tape recorder over the record player as a music machine could be made clear to one and all.

"My God, here's this company which is the latest word in both audio and video! We can make a killing!" Such was the attitude (as paraphrased by a subordinate) that Roberts brought to Ampex.

Like any new boss, he took a hard look at the projects he had inherited from the old boss, and when he looked at the Ampex-Sony project, he was appalled. What was Ampex doing, he wanted to know, giving away its technology to another company—and a Japanese company at that? To Long and Gundy, the nonbroadcast market for VTRs—the market they had been willing to let Sony enter—was not terribly important. But Roberts envisioned Ampex as a consumer company. He foresaw a time when its video as well as audio recorders would make their way into millions of ordinary homes. (It might not be anytime soon, though, to judge from the first VTR that Ampex, under the Roberts rule, launched into the consumer market—a helical-scanning machine which was built into a "complete home entertainment system" in an all-hardwood cabinet and offered for sale at a mere thirty thousand dollars in the Christmas 1963 Neiman-Marcus catalog. Officially christened the Signature V, this staggering concoction was informally dubbed Grant's Tomb after Gus Grant, the marketing manager who conceived it. The nickname was especially appropriate since, as one Ampex engineer explained, "It was about the right size.")

Soon after Roberts's arrival, Roizen returned from Tokyo, carrying the first transistorized circuits, which he said years later, "were put in a cupboard and never installed in a machine and are probably gathering dust still." Roberts was determined to do nothing that could be interpreted as a sign of approval of the agreement with Sony. Indeed, it was his position—and the position of Ampex's lawyers—that the one-page document signed by Gundy, Morita, and Ibuka was not an agreement at all, just a preliminary memo.

Sony thought otherwise. As a Japanese company it was used to doing business on the strength of brief and unlawyerly documents, and it made its attitude toward this one perfectly clear in 1961 by putting a VTR on the market—a transistorized, transverse-scanning machine known as the SV-201.

Roberts and his people were livid, and they made their feelings known. By now, however, Japan had become a source of many headaches for Ampex. After the initial rush of orders the company's attempts to sell VTRs to Japanese broadcasters had gone nowhere. Certain official approvals were necessary before a foreign firm could do business in the country, and they were not forthcoming. Ampex's Japanese patent applications also went into a holding pattern, while two domestic companies filed applications of their own, which seemed to imply an attempt to contest Ampex's legal position. More ominously, several Japanese firms (not including Sony) began to supply broadcasters with VTRs that closely resembled Ampex's—without, of course, obtaining Ampex's permission. One company, Shibaden, had the audacity to put out a machine that was a dead ringer for an Ampex VTR—down to the useless holes in the top plate, which Ampex had put there by mistake. When Ampex's people complained, MITI officials unapologetically told them that they had better take on a full-fledged Japanese partner if they hoped to sell any VTRs in Japan beyond the fifty or sixty they had sold to date. And to end what they regarded as out-and-out patent infringement, they would have to negotiate licenses with the alleged infringers, allowing them to go legitimate.

Having no real choice in the matter, Ampex yielded on both points, and it proceeded to set up a joint venture with Toshiba. Toshiba would hold fifty-one percent of the stock, and Ampex forty-nine percent, in a company that would make and sell VTRs for the Japanese market. (In fact, Ampex shipped the important components over from the States, leaving the new entity, Toamco, to do the final assembly and add knobs and other decorative accoutrements. It was a kind of arrangement that would become more common between Japanese and American companies in the seventies and eighties—only with the roles reversed.) As soon as Ampex began playing by Japanese rules, its various applications moved routinely through the bureaucracy, and the Toamco VTRs quickly captured a handsome share of the Japanese broadcast market. The company also realized healthy revenues from patent contracts with the reformed infringers. Shibaden was even good enough to pay for past as well as future acts.

But Sony refused to be lumped into the same category with these miscreants. It continued to assert its right to make nonbroadcast VTRs without paying a royalty, and though its first attempt, the SV-201, was too big and costly for the customers Sony hoped to attract, a series of smaller and less expensive models followed: the PV-100, a transistorized, helical-scanning machine which was a fiftieth the size of the original Ampex VTR (and, to add insult to injury, was soon adopted by American and Pan American airlines for in-flight entertainment); the EV series, a smaller system that used one-inch-wide instead of two-inch-wide tape and was meant for use in schools (the initials stood for educational video); and finally, in 1965, the half-inch CV-2000 (for consumer video), ballyhooed as the world's first "home videocorder" using "the newest major breakthrough in magnetic recording techniques, compact in size and low in cost."

The "newest major breakthrough" was Kihara's notion of skip-field recording. For convenience's sake, one may imagine a television picture as consisting of frames that change thirty times a second, compared to the frames of a motion picture which change twenty-four times a second. But on a movie screen the whole frame appears at once, whereas on a TV screen the frame is really a succession of lines illuminated by an electron gun, one by one, in a left-to-right, top-to-bottom order which resembles the way people read a page in a book. To keep this aspect of the process invisible to the viewer, the electron gun scans alternating lines—first the odd ones, then the even (or vice versa)—and each sequence is called a field. The CV-2000 (and its successor CV models, the 2010 and 2020) recorded only one of the two fields in a frame and reproduced it twice.

The drawback of skip-field recording was a slightly hazy picture, but TV was an inexact medium in which "ghosts" and other irregularities were liberally tolerated by most viewers. The virture of skip-field recording was a low rate of tape consumption—half that of full-frame recording—and the resulting ability to build a more compact machine. A seven-inch reel of half-inch CV tape could hold an hour of material—either original programming shot with the six-pound camera that came as an accessory or ordinary TV shows recorded off the air. The first two models—both with built-in TV sets—cost $1,000 (for the so-called portable version, which weighed sixty-six pounds and came in a Leatherette case) and $1,250 (for the living-room version, which weighed seventy pounds and came in a walnut cabinet "manufactured in the United States," as Sony proudly noted in its press release). The more expensive model even had a timer, "to record a television program while the owner is away from home."

From the beginning Sony had looked on the VTR as a potential consumer product. Morita regarded the CV machines as a declaration of independence against the tyranny of time. "People do not have to read a book when it's delivered," he liked to say. "Why should they have to see a TV program when it's delivered?"

In both Japan and the United States customers were put off by the complexity as well as the price of the CV machines. Tape in its naked state, wound around an open reel, was tricky to thread and all too likely to break, twist, tangle, or spill across the floor. And the CV-2000 had the additional misfortune of being a black-and-white machine which came along just when people were starting to develop a fondness for color. Nevertheless, the CV models enjoyed respectable sales to educational and industrial customers—Sony adroitly reassigned the initials to stand for "commercial" instead of "consumer" video—and they may have been the last straw as far as Ampex was concerned.

In 1966, Ampex formally renounced the patent exchange, demanded an eight-percent royalty on Sony VTRs, and—when Sony refused to pay—sued for patent infringement in the United States. Sony countersued for breach of contract, and two years passed before the companies reached an out-of-court settlement. They went at it in the American marketplace, too, with each company trying to sell one-inch helical-scanning VTRs to semiprofessional, industrial, and educational customers. Ampex fared well at first, but Sony gradually made inroads, especially in the educational market. As a professional company Ampex was used to dealing with knowledgeable buyers like the audiovisual (AV) directors who made the purchasing decisions for schools and school systems. Some of them had backgrounds in broadcasting, and they were plenty particular about the features a VTR should include but not so concerned about ease of use. "They would ask questions such as 'What about more of this or more of that?' " recalled Koichi Tsunoda, a Sony executive who spent a good part of the sixties selling VTRS in the United States. "The Ampex salesmen were very amenable to customer requirements. They said, 'Okay, we'll try to improve this or improve that.' Gradually their price was going up and up. Sony, on the other hand, felt that the home market would finally be the biggest market, so we tried to make a simpler machine for the teachers and the students."

Despite the AV directors' taste for Ampex's machines, Tsunoda noticed that they tended to languish in storage rooms once they had been purchased. Teachers found them intimidating. Teachers found the whole idea of audiovisual education intimidating. Futurologist types had begun to insinuate themselves into school systems, and they envisioned closed-circuit TV networks that would bring instruction to large numbers of students at once. "Teachers were very reluctant to use videotape recorders because they felt 'Machines are eventually going to replace me,' " Tsunoda recalled. "We said video could be used not only in AV room or broadcast station of educational institution. The key to success in educational market was immediate playback, for role-playing." With its simpler machine, and with the lightweight video camera which it introduced as an accessory, Sony was in a better position to fill this bill. "We wanted the student and the teacher to use it," Tsunoda said, "and, if possible, to bypass the AV department."

Ampex's one-inch models were heavier than Sony's, and the tape guide system was erratic, so that programs recorded on one machine couldn't always be played back on another. "People who actually were using videotape recorders felt Sony machine didn't have all the goodies—the bells and whistles—but it was more reliable and more convenient," Tsunoda said. "So gradually we are replacing Ampex."

Richard O'Brion, a Sonam sales rep in those years, summed up the Ampex one-inch machine as a "disaster" and explained, "We were able to cut them to ribbons because the damn thing didn't work. They'd sell them and take them back again and again. They'd sell the same machine two and three times." Ampex's traditional customers were TV stations. If a problem developed, they had maintenance people to tend to it. But a cantankerous machine was a dilemma in a school setting. If the school treated it as a delicate and precious piece of equipment which had to remain in the AV director's custody, it might not be used very much; otherwise, it might break down.

Ampex was having other troubles in the late sixties—troubles unrelated to Sony. As the prerecorded tape business grew, Ampex's place in it shrank. The *Wall Street Journal* analyzed the company's predicament under the headline "The Music Stopped: How Ampex Saturated Recorded Tape Market and Got Soaked Itself." In 1972 William Roberts had to resign as chairman after the company announced unprecedented losses. With Roberts at the helm, according to a "knowledgeable source" quoted in the *Journal,* "Ampex tried to do more things than it had money for." After his departure the company spent the better part of a decade finding its identity again.

The invention of the first commercially successful audiocassette by Phillips in 1962 set many minds to thinking about putting videotape into a container. In his sales pitches to American schoolteachers Tsunoda noticed that they shied away from threading a reel of tape, even with a Sony salesman standing by. "I'm afraid to do it—why don't you do it?" they would say. Tsunoda was inspired to send a memo to Tokyo in 1964, urging a cassette-format VTR to overcome this sort of resistance. Ibuka was also high on the idea. The audiocassette, he liked to point out, had vastly expanded the market for tape recorders, turning them into a product for general consumers as well as professionals. Why were there no videocassettes? he kept asking.

The answer to his question was rooted in the original challenge of videotape recording—the intricacy of the signal and the large quantity of tape required to handle a reasonable length of programming. By the mid-sixties black-and-white VTRs were giving way to color, which complicated things. Now there was the color, or chrominance, signal as well as the black-and-white, or luminance, signal to deal with; and with performance standards on the rise, the skip-field technique no longer struck Kihara as an acceptable way to go. With full-frame recording, however, it was barely possible at the time to fit an hour of color programming onto an eight-inch-diameter reel of one-inch tape. A cassette capable of holding two such reels—for supply and take-up—would be prohibitively large. And it would be no small feat, what with revolving heads and helical scanning, to construct a machine capable of loading the tape without human intervention.

In the face of these difficulties, Kihara toyed with a half measure—a cartridge format. A cartridge is a container for the supply reel only; the take-up reel remains a separate unit, and the tape must be rewound before the cartridge is removed. Along with some sort of automatic or semiautomatic threading system, however, a cartridge might theoretically serve the same purpose as a cassette, eliminating contact between hand and tape; and the Electronics Industry Association of Japan (EIAJ), which was trying to develop a standard format for nonbroadcast VTRs, seemed to regard the cartridge concept as promising. Kihara's experience with cartridges went back to the Babycorder, the battery-operated audiotape that Sony had sold in the fifties. But none of roughly a dozen cartridge prototypes that came out of his shop in 1968 and 1969 satisfied the management, and the cartridge model VTRs put out by other Japanese companies met a cool reception.

While Kihara was moving from cassettes to cartridges and back again, Sony was bringing together a number of complementary developments that were to contribute to more space efficient video recording. Iron oxide gave way to finer, more sensitive magnetic powders on the tape, while new materials and advances in micromanufacturing paved the way to a smaller recording head. Ibuka had issued a call for a book-size cassette, and Kihara and his team came up with one in the fall of 1969, using three-quarter-inch-wide tape. Then they licked the loading problem with a system called U-loading, in which a length of tape is automatically pulled from the cassette by a series of pins mounted on a ring and is wrapped in the shape of a sideways U around the drum. "At first glance it looked a little complicated," Kihara said, "but

Mr. Ibuka decided that it was the best compromise—the best system to serve consumer needs."

The U-matic, as Sony called its machine, was the first videocassette recorder to reach the market. Kihara and his superiors saw it as the culmination of a long search. As Tsunoda recalled, "This time we had nice color and a nice cassette. So this time we thought, 'All the fear is gone. Now we can come into the consumer market.' "

But first Sony took an unprecedented step. As excited as Ibuka and Morita were about the new product, they decided to show it to two of their Japanese competitors before putting it into production, with the thought that all three companies—Sony, Matsushita Electric, and JVC (a company largely owned by Matsushita, although independently run)—might adopt the U-matic as a common standard. This was not a move that came naturally to Sony. In the fifties it had become large and successful by maintaining monopoly control over the Japanese tape recorder business. At industry gatherings in the sixties—when the Electronics Industry Association of Japan had been trying to foster cooperation in VTR design—Sony's representative had been given to asserting the superiority of their ideas to those of other companies, and the meetings had sometimes degenerated into confrontations between Sony, on one side, and everybody else, on the other.

But Sony was not the only company trying to come up with a compact, simple color video recorder. Matsushita, JVC, Toshiba, and Hitachi all had such products in the works, and there would be formidable competition to face in the United States and Europe, too. A proliferation of incompatible formats, Morita and Ibuka feared, could cause consumers to recoil in confusion or even to embrace an inferior system simply because one large company was promoting it. The EIAJ had been pushing hard for compatibility, and the management of Sony decided that it might be an idea whose time had come. After all, Monta and Ibuka reflected, when Sony had reluctantly released its audiotape recorder patents in the fifties, its share of the market in Japan had fallen from ninety to thirty percent, but the size of the market had grown by a factor of ten, so the company had gained, not lost, by sharing its technology.

Early in 1970, Sony showed its videocassette and videocassette recorder—along with technical specifications and test results—to representatives of Matsushita and JVC. After weeks of contemplation they responded favorably. They would adopt the Sony format, they said, if Sony would accept a few modifications, including the use of a color-recording technique developed by JVC; an increase in the size of the head drum, to simplify production; and in increase in the space between the tracks on the tape, to reduce the risk of tracking error. Sony, according to Kihara, "answered decisively that we would compromise."

When the first Sony U-matics went on the market in 1971, they met an ambivalent reception. Ordinary consumers in Japan and the United States found both the machine and the cassette too big and too expensive. For a variety of reasons, a product originally intended to sell for five hundred dollars wound up costing well over a thousand dollars in its simplest version, and the cassettes cost thirty dollars each for an hour of recording time. But the U-matic aroused surprisingly strong interest from educational, industrial, and even professional customers, and Sony quickly proceeded to modify the machine and to supply accessories, such as an editor-player, with those markets in mind. In 1973 the American company Consolidated Video Systems developed an inexpensive, digital time-base corrector—a computer that compensated for the tracking errors that had plagued helical-scanning VTRs since they had been invented. The performance of the U-matic, already good, got better, and before long it invaded the broadcasting world. Transverse-scanning VTRs were extremely bulky for use outside the studio, so film rather than videotape was still the preferred medium for news and documentaries (although an Ampex backpack-style machine had been used to cover the Mexico City Olympics of 1968, among other events). But in June 1974, over the objections of its film people, CBS-TV News used a U-matic and one of the lightweight, high-quality video cameras produced by another Japanese company, Ikegami, to report on Richard Nixon's trip to Moscow. Electronic news gathering, or ENG, had been born. Videotape cost less than film to use, and it made editing a matter of pushing buttons instead of cutting and splicing. Decisively rejected by the market for which it had been intended, the U-matic became a stunning success just the same.

LASERS AND LUNCHES

THE NEWSPAPERS AND newsmagazines of the late 1960s and early 1970s did their best to prepare the American people for the onslaught of something like the home videocassette recorder. Soon technology would make it possible to "ignore commercial TV's rigid timetable and standardized fare," *Life* magazine predicted in October 1970. The *Life* correspondent confessed to some uncertainty about the nature of the machine that would work this miracle. Although Ampex's four-head, transverse-scanning VTR remained the standard for broadcast use, the competition to produce something similar for the home was wide open. The winning entry might be a variant of a record player, a movie projector, or a tape recorder, or it might be like nothing that had come before. Whatever, its "impact on America's viewing habits and life-styles" would be "greater than anything since the advent of television itself."

Only one important feature of the home video world as we know it could never have been imagined from the forecasts of fifteen and twenty years ago: that the ambitious plans of more than a dozen American companies—the likes of CBS, RCA, Avco, Magnavox, Motorola, Kodak, Bell & Howell, Fairchild, Zenith, MCA, and Polaroid—would go up in smoke, and that when the smoke cleared, the Japanese would control the field.

One could never have guessed it, certainly, from the mesmerizing oratory of Peter Goldmark, the Hungarian-born inventor of the long-playing record. As president of CBS Labs Goldmark devoted half a decade of his life and thirty-three million dollars of his employer's money to something called electronic video recording—EVR—which, depending on the metaphor of the moment, was the "video long-playing record of the future" or the "greatest revolution in communications since the book." Goldmark had dismissed magnetic tape as far too expensive a medium for his concept, which, despite the name, would be a playback-only system.

> It seemed to me [he wrote in his autobiography, *Maverick Inventor*] that the answer had to lie in film, which could hold more information per frame than a similar square of video tape and hence would be cheaper to make. However, I knew we would have to develop the tiniest frames every produced. In that instant I saw the outlines of a system in which miniaturized film could be played through the TV set. I also saw that every home could use this miniaturized film cassette. Great libraries of special film containing much of the world's information would be available to anyone, just as library books are now available across the country.

When Goldmark made his first pitch for development money in 1961, William S. Paley, the chairman of CBS, turned him down flat. Paley, according to Goldmark, found the idea too threatening to conventional broadcasting. (Paley's memoirs suggest another explanation: that after a disastrous attempt, at Goldmark's urging, to turn CBS into a manufacturer of TV sets, Paley was in no mood for another foray of that kind.) Goldmark was only momentarily deterred. He took his project to the federal government—a traditional recourse for the frustrated corporate inventor—and, in the name of national security (the magic words *weapons training* were mentioned), persuaded the United States Air Force to give him some R&D money on the sly. "Washington has a way," Goldmark wrote, "of supporting research-and-development projects by raising the overhead fee on old projects and allowing the increase to be pumped into new projects"—an arrangement, he added, that, "while tricky at first glance, is good both for the government and for industry because the government becomes the instrument of innovation—something I am afraid industry often is not."

By 1964 EVR had the support of Frank Stanton, CBS's president (and Goldmark's patron), and the engineers at CBS Labs had brought the system to a state of readiness for the eyes of Paley. But the eyes of Paley were not ready. Motivated, according to Goldmark, by "a feeling . . . just short of irrational, against anything that appeared to threaten home television," Paley refused to attend the demonstration that had been arranged for him. "How could the head of a large, responsible organization behave this way?" Goldmark asked.

For a time he lost hope. Then he had a talk with Stanton. Paley had no objection to the technology per se, Stanton explained. He simply considered it too expensive for a consumer product (a point on which Paley's judgment would stand the test of time). Stanton advised Goldmark to promote EVR as an educational tool instead and to strictly avoid mention of any applications in the home.

"No home," Goldmark obediently replied. "Absolutely not."

As he soon discovered, however, he had only to describe the general capabilities of the system, and others could be counted on to speak of the forbidden. By the summer of 1967 CBS was showing a prototype that, with a seven-inch-diameter cartridge of 8.75-millimeter film, played an hour of sharp-resolution black-and-white programming through a TV set. (In short order it would be able to play a half hour of color, Goldmark declared.) How could any self-respecting journalist fail to see the wider implications, especially when Goldmark predicted that EVR players would be sold for as little as two hundred dollars and the film cartridges for as little as seven?

"Soon You'll collect TV Reels, Like LP's," was the headline over a *New York Times* story about EVR on September 3, 1967.

In March 1970 CBS held a public demonstration at the Hotel Pierre in New York City. The audience broke into applause when it saw a color picture that one member of the trade press called "better than this reporter has ever seen on a TV set before." By then CBS had created an international consortium, the EVR Partnership, to sell players and cartridges around the world, and an impressive roster of companies had sworn their allegiance to the system in one way or another. Motorola would be turning out the EVR players, at eight hundred dollars each. Twentieth Century-Fox had agreed to license a library of films for release on EVR cartridges. Equitable Life Assurance was going to use EVR as a training medium for its sales agents. The momentum seemed unstoppable. Hardly a leading prognosticator could be found who did not believe that an important new industry was about to be born.

"There wasn't a week that went by that we weren't in *Variety*, *Business Week*, *Life*, or some other important publications," Robert Brockway, the president of CBS's EVR subsidiary, said later. "The interest was tremendous. From a marketing standpoint, it had complete acceptance."

Next to Goldmark himself, EVR probably had no greater enthusiast than the TV critic Jack Gould of the *New York Times*, who took a machine home with him after the Pierre demonstration and proceeded to wax ecstatic over the "captivating delicacy" of the "soft and true pastels" in an excerpt from the movie *The Prime of Miss Jean Brodie.* Gould concluded that "the implications and possible applications border on the staggering." Fortuitously, his employer was equally enthusiastic. The *Times* had signed on as the first independent producer of EVR programming, and it had fifty educational programs in the works.

By November 1970, when Motorola delivered the first color player to Equitable Assurance, the taboo on references to the home had been lifted. A CBS press release predicted that "in 1972, as player penetration expands and cassette production grows, EVR will become a mass-consumer product. Within a few years, EVR players will be commonplace in the American home. All this will happen faster than you might think. In this electronic age, growth is geometrical not arithmetical."

Not even the most delirious publicist could have known just how swiftly EVR's fate would be decided. For as long as they humanly could, Brockway and the other EVR executives averted their gaze from the production process and threw themselves with utter compartmental-mindedness into the task of promotion. But there comes a time in the life of any business when people begin to expect the marketing effort to be accompanied by tangible signs of economic life such as production, distribution, sales, and cash flow. When these elements failed to materialize for EVR, the management of CBS started asking questions.

Among the relatively small circle of people who actually knew how EVR worked and how things were going at Goldmark's shop in Stamford, Connecticut, and at the cartridge-making plant in Rockleigh, New Jersey, support for EVR turned out to be good deal less fervent than in the world at large. The central problem, well shrouded from outside scrutiny, involved the process of generating the masters from which EVR cartridges had to be struck. This was done by exposing the film in a vacuum chamber to a beam of electrons, and the machine that did the job went by the rakish name *electron beam recorder*. It sounded gloriously futuristic. Unfortunately it was all too futuristic. In the present, as far as anyone could tell from the doings in Stamford, where the world's first two electron beam recorders were housed, it took about two and a half years and two million dollars, minimum, to build one, and when finished, they produced masters of extremely erratic quality. What's more—and this was the "narrow end of the funnel," in Brockway's words—the masters had to be struck in real time; the master for a half hour program, in other words, took half an hour to make.

Harry Smith, an EVR vice-president who assumed responsibility for manufacturing in the final months of the enterprise, made some grim judgments. To meet the needs of the first EVR customers—corporations and schools, mostly—CBS would have to produce a wide assortment of programs, but relatively few copies of each. These short production runs had brutal cost implications; a single twenty-minute EVR cartridge, Smith calculated, would cost between fifty and a hundred dollars to produce. About the time he arrived at this estimate, Sony came out with the U-matic, with hourlong cassettes which cost thirty dollars and could be duplicated anywhere at an insignificant expense. Smith concluded that EVR was economically untenable. When he laid out the data to his superiors, they could not but agree, and in August 1971—just as "EVR was on the brink of spawning a new series of industries," according to Goldmark—CBS bailed out. For all the publicity, it had produced only a few hundred of the precious cartridges.

"Paley fought it, almost as though possessed of a death wish," Goldmark wrote in his memoirs, summing up the difficulties that did EVR in. Paley, in *his* memoirs, wrote that "I began to look upon Peter

Goldmark . . . as a thorn in my side. That year, he turned sixty-five and retired from the company."

RCA was also busy marketing a product that did not entirely exist. The RCA system employed lasers *and* holograms—what could be more futuristic than that?—to record a picture on a piece of cheap plastic film that could be played back through a TV set. (As with EVR, the recording would happen at the factory, not in the home.) It was never anything more than a distantly promising experiment, according to George Brown, who served as RCA's vice-president for research and engineering in the sixties. All the same, it captured the imagination of David Sarnoff's son Robert, the reigning president of RCA, and Chase Morsey, his vice-president for marketing.

In 1969 Morsey decided to go public with the hologram system. "The first action," Brown recalled, "was the coining of acronyms. The holographic system was first called HoloTape and soon was changed to SelectaVision. PREVS for 'prerecorded electronic video systems' was an all-encompassing acronym for SelectaVision and for any future method yet to be invented. After a few years . . . SelectaVision was announced to be the generic term for any home player system since the name was considered too good to be lost." Morsey had spent the 1950s at the Ford Motor Company, and one night, at a slightly rowdy departmental dinner, the engineers coined another, unofficial acronym: EDSEL, for "electronic devices and systems for education and learning."

That September a throng of reporters and editors descended on the David Sarnoff Research Center for a look at the prototype—the picture "ranged between 'poor and lousy,' " according to Brown's account—followed by an elaborate lunch in a tent pitched outside the lab, and a few words from Sarnoff and Morsey about the impact on life in the latter part of the twentieth century. With considerable precision for such a far-off event, Morsey announced that the player units would go into production in 1972 at a price in the neighborhood of four hundred dollars, while the half-hour-long "program albums" would sell for about ten dollars—far less than videotape or what was referred to as "proposed film replay media," which meant CBS's EVR. Sarnoff proudly declared HoloTape to be "the most thoroughly market-researched development ever to emerge from the company's laboratories." When the afternoon group departed, a second sitting of journalists replaced them, and the performance began anew.

To be fully understood, the furious promotion of HoloTape, like that of EVR, probably must be viewed in the context of a long corporate rivalry between CBS and RCA. Neither company's management could abide the thought of conceding this glorious new market to the other. Both might have done better, of course, to have applied their competitive energies to products that could actually be manufactured and stood a chance of success in the market. But lacking anything of that kind, they fought with the best weapons at hand: the acronym, the press release, and the demonstration lunch.

With CBS and RCA battling for the lead in column inches of press coverage, it was hard for anyone else to be taken seriously. Still, plenty of companies tried, and some big names among them. Ampex had a half-inch "auto-threading" VTR aimed at the consumer market. At first it went by the name InstaVision, then it was InstaVideo, and then it was gone (concurrently with William Roberts as Ampex's chief executive). Kodak briefly marketed a twelve-hundred-dollar black box which allowed super-eight movies to be played through a TV set. And in December 1972, at an international conference on such technologies in Cannes—a "meeting of a non-industry," as *Television Digest* called it—MCA gave the first public exhibition of its laserdisc system. Regrettably, its prototype was too precious a cargo to undergo such a long journey, MCA had to explain, so instead of delivering an actual disc and disc player, the company had been obliged to stage the demonstration in California and send a Sony U-matic recording of it to France—a bad omen if ever there was one.

In later years, when the veterans of EVR, HoloTape, and the other evanescent wonders of the pre-Betamax era fell to reminiscing, they tended to be strongly critical of those projects in which they did not happen to have been personally involved, and no entry was viewed with more general scorn than Cartrivision, a system developed entirely outside the normal channels of the TV and consumer electronics industries by a group of New Yorkers whose major business credential was having had the great good sense, in the early 1950s, to import the Volkswagen Beetle to the United States.

Cartrivision was not a public relations triumph. Its promoters obviously lacked the talent for concealment that was so highly cultivated at CBS and RCA. When they composed a press release that was at variance with the actual state of their research, the press, for some reason, seemed to root out the truth almost immediately. After one early demonstration *Television Digest* commented on the "lack of crispness, color drop-outs, and blurred motion" of the system and added: "Our talks with CTV marketing and engineering executive left us with the strong impression that everything is not quite as resolved as Cartrivision's show presentation would indicate. Machine itself still has long engineering road ahead." When the promoters put out a stock offering, a columnist in the *Wall Street Journal* seized on it as a case study in the hazards of "buying 'dreams.' "

And yet, of all the systems conceived, announced, and demonstrated in the pre-Betamax years, Cartrivision was the only one of American manufacture to be submitted to the judgment of the marketplace, and after that judgment had been rendered, Cartrivision's promoters could claim with reasonable fidelity to the truth that fully three years before Sony unleashed the home video revolution, another company—an American company—had been selling a product capable of doing everything a Betamax could do.

"We had color, we had liftoff, we had roll 'em," said Frank Stanton, the "other Frank Stanton," as he is known in the halls of CBS—the Frank Stanton, that is, who ran Cartridge Television, Inc., the company that gave the world Cartrivision. "It was a crude cassette at first," Stanton recalled. "We had a lot of color dropouts, but basically it worked just fine."

"It was not dissimilar to today's hardware," Lawrence Hilford, another Cartrivision veteran, recalled. "You dropped the cassette in a slot, pushed a button, and it played."

Like the Betamax, Cartrivision was a videocassette recorder that came equipped to record programs off the air, although in its advertising the company treated that feature as only one of many. The cassette, which used half-inch tape, was slightly smaller than the U-matic's but square rather than rectangular, because the supply tape and the take-up were wound around a common core. this odd geometry and the decision to use the skip-field recording technique of the Sony CV-series machines made it possible to put almost two hours on a single cassette. That achievement, in turn, was basic to the most novel idea of Cartrivision: that the machine and a wide assortment of prerecorded programs—above all, feature films—should be marketed in concert.

Akio Morita had made overtures to a few American movie companies about putting their works on U-matic cassettes, but he had been unable to overcome the studies' resistance to letting the ownership of even one copy of a movie pass out of their hands or the studios' desire, in any form of distribution, to know how many times each copy of a film was viewed. The Cartrivision people went out of their way to answer these concerns. Their movie cassettes would come in a special red box, with a locking device which made them impossible to rewind in the home. Only retailers would have access to the special rewinding machine, which would be equipped with a counter so that they, too, could engage in no shenanigans without the studio's knowledge. And the movies would be available for rental only.

"What you never do in this business is sell a film," the other Frank Stanton explained. "Where did we learn that? From the movie companies! When they distribute a film, they lend a print to the theater and participate in the profits. The theater never owns anything."

The Cartrivision sales pitch went over nicely in Hollywood. Several of the studios agreed to make their films available, and Columbia Pictures became a partner in the Cartridge Rental Network, a company set up to handle the distribution of movies on Cartrivision cassettes. The management of Sears, Roebuck took a liking to the venture and agreed to help launch it at the retail level. Stanton and his partners also persuaded Avco, a major defense contractor, to invest millions of dollars in turning a former weapons plant in Richmond, Indiana, into a facility for manufacturing recorder-player units, and in building a cassette and recording head factory in San Jose, California. Later some twenty million dollars were raised from tens of thousands of ordinary investors who bought out a Cartrivision stock offering, undeterred by the skeptical coverage in the *Wall Street Journal*.

The first Cartrivisions went on sale at eighteen Chicago area Sears stores in June 1972. "From dream to product in two years," Stanton announced at a champagne party marking the occasion. The initial catalog of black cassettes (for sale only) contained 111 titles, including *Fishing with Gadabout Gaddis, Rembrandt and the Bible, Erica Wilson's Basic Crewel,* and Chekhov's *The Swan Song,* billed (although Chekhov might have been surprised to hear it) as "the first dramatic work made expressly for cartridge television." The catalog of red cassettes (rental only) had 200 titles, including *Casablanca, It Happened One Night, Red River,* and *Dr. Strangelove.*

The Chicago newspapers ran double-spread ads for "a complete home entertainment and learning center for you." (Cartrivision came only in a console model, with the recorder-player unit annexed to a color TV.) "First of all," the ad said, "Cartrivision is a color television set which will provide you with fine color reception . . . but Cartrivision goes television, and any other entertainment media, one better." It "not only lets you watch what is playing on television on a particular evening, it lets you watch the program even if you aren't there . . ." It was "an instant home movie maker" with "its own camera featuring Instant Replay, and a portable microphone." And "whenever you want," the ad promised, "you can watch movies—Hollywood, foreign, or Underground—music, dance, sports, poetry . . . Take a golf or a tennis lesson from your favorite pro. Watch a great basketball game, a baseball game, a football game—hockey, soccer or golf. Let the greatest actors in the world tell you a story. Take a painting lesson from a master, a piano class from a virtuoso . . . Learn how to get thin, how to get fat, how to exercise, how to meditate. Watch the world's chefs show you how to flambé a steak."

Like many complicated gadgets down through the years, Cartrivision was described as "so easy to do, your child can operate it." But as Groucho Marx observed in *Duck Soup,* "A four-year-old child could understand this . . . Run out and find me a four-year-old child. I can't make head or tails out of it." The salespeople at Sears found Cartrivision a mouthful to explain. As Hilford recalled, "They knew how to talk about furniture. They could tell you how to turn a TV set on and off. But they had never run one of these things. They hadn't been trained, and at some of the stores the programming was handled on another floor, or if you wanted to rent something, there was a system where you could call and get something a few days later from UPS and you'd have to return it the same way. It was a mess."

"The salesman tended to treat it like a giant TV set," another Cartrivision executive recalled, "and this was when giant TV sets with heavy wooden cabinets were going out of style. Also, a lot of them were just in awe of having to sell a sixteen-hundred-dollar item."

Some sales were lost because the would-be buyers couldn't meet Sears's credit requirements, and the Cartrivision people belatedly realized that Sears had been the wrong place to start selling such a costly and sophisticated product. But the complaints ran both ways. The retailers complained of difficulty in obtaining the advertised programs—especially the feature films. Potential customers were holding back, they said, because of lack of confidence in the availability of the prerecorded cassettes that had been so hotly promoted. And in retrospect, it struck some people that the customers most likely to buy Cartrivision were those most likely to own a color TV already. Many a shopper was heard to say: "Maybe I'll buy one of these when my present set wears out." Eventually the company intended to market a separate recorder-player deck—but eventually proved to be too late.

The performance of the machine itself—and the skip-field recording process it employed—did not escape criticism. "It was a jumpy picture," David Lachenbruch of *Television Digest* recalled. "If you looked at it carefully, you could see it jumping, but most people were so impressed that they got a picture at all they didn't notice."

The no-rewind feature of the red cassettes, on the other hand, performed all too well. If a screaming infant summoned a viewer away from his Cartrivision set in a hurry and made him miss a few minutes of a movie, there was no way to recapture those minutes without a return trip to the store—and no way for the store to rewind the cassette without reporting a fresh transaction to the Cartridge Rental Network. (The unrewindable cassette might have been a rash idea, Stanton later admitted; the studios, he said, had made it a condition of doing busines with him.)

But all other failings faded in insignificance in November 1972, when, with Christmas just around the corner, Cartrivision was hit by a catastrophe that seemed to have been lifted from the Alec Guinness movie *The Man in the White Suit.* The tape sitting in warehouses and stores all over the country—blank as well as prerecorded—spontaneously began to decompose. "Almost our total inventory proved to be defective," Hilford said. "and when one of those tapes was played on the machine, it could destroy the heads. So we had to recall virtually every tape in the country, and then we had this wonderful machine out there with nothing to play on it." After announcing plans to produce up to 50,000 Cartrivision units in 1972, the company had to admit in early 1973 that only twenty-five hundred had been sold.

In March, with the tape problem coming under control, Cartrivision launched what *Television Digest* called a "do-or-die plan to bring . . . hardware, software, ad & promotional efforts together for concerted drive on West Coast."

"We had to prove that the machines worked and that people would buy them and rent the cassettes with sufficient frequency to get Sears to break out nationally and to get everybody else to come on line," Hilford said. "So we set up something like a hundred-and-thirty-store test market in California—from San Francisco on down—involving Sears and Montgomery Ward and Macy's and a bunch of smaller appliance stores. And the rule was that no store could go on line unless we had trained the salespeople in demonstrating the hardware and unless a sufficient supply of cassettes was available and visible with the hardware." Avco agreed to help out with four-year consumer loans, and free movie rentals would be given away with each machine.

"Well, after about eight weeks the test was going gangbusters," Hilford said. "The numbers showed that people were renting at the rate of something like two-point-eight units a week. The hardware was selling very nicely. Sears was saying that the repair rate was equivalent to the third year of color television, which was not sensational but eminently satisfactory, and Sears told Avco that on the basis of the test they were ready to break out nationally."

During those same weeks, however, outsiders and insiders alike began to take a worried look at the formidable gap between expenses and earnings. Cartrivision had been conceived on a grand scale—as an all-out assault on the American way of watching TV—and it had a budget to match. To prepare for heavy production a few years down the road, the company had invested in an expensive high-speed tape duplication system. Handling the low volumes of the moment with this state-of-the-art machinery was like taking a trip across town on the space shuttle. Huge sums of money had gone into the effort to launch a nationwide marketing operation and to spread "neighborhood cartridge libraries" across the land. To administer it all, the company had hired three hundred employees to man its headquarters in Palo Alto, and there were handsome offices in New York as well, for Stanton belonged to a school of marketing theory which stressed the importance of that intangible a "class operation"—something that, in the execution, always seemed to entail tangible things like carpeting on the wall. And of course, there had been a number of unexpected expenses—notably the tape recall and a patent suit filed by Ampex, which had led to an out-of-court settlement that would cost Cartrivision a minimum of three hundred thousand dollars a year in license fees.

When Wall Street looked at all these obligations, in one column, and at the paltry revenues, in the next, it lowered its opinion of Cartrivision's stock from the original twenty-dollar-a-share price to nine dollars. When the directors of Avco looked, they began to question the thirty or forty million dollars they had already put up and the eight million more that Avco's chairman James Kerr seemed disposed to add to the kitty.

Sensing that Avco's largess had reached its limit, Stanton embarked on a drastic cost-cutting effort, closing offices, laying off workers, and postponing plans for new models—an eminently logical program in the circumstances but one that went completely against the spirit of everything that had come before and cast further doubt on the basic sanity of the enterprise. When Cartrivision went into its retrenchment phase, the vultures went into their landing pattern. Wall Street downgraded the stock again, to less than there dollars a share, and in June—"even as massive California effort was beginning to show . . . some progress," according to the relatively dispassionate view of *Television Digest*—the board of Avco voted to pull the plug, writing off an investment that had grown to forty-eight million dollars. A few days later

Stanton and his colleagues filed for bankruptcy, issuing a terse announcement that operations to date had "substantially exhausted the funds available."

For months before and after the bankruptcy filing, Stanton scoured the corporate horizon for additional funds. "I still thought it was worth saving," he said later, speaking from the dry ground of the mid-eighties, after he had returned to a prosperous career as an importer of cars and a developer of real estate. "As the years go by, I think so more than ever. We approached a number of companies, and they could have walked in and made a marvelous deal. Had we been able to continue, we would dominate the industry today." Ten million dollars would have made all the difference, by Stanton's reckoning, and it seemed to him that "if you're worrying about ten, twenty, or thirty million dollars, you shouldn't be in business." Unfortunately all the data he had to show his would-be rescuers, apart from those eight weeks of improved sales figures in California, were negative, and it was not clear how Cartrivision would conquer what *Television Digest* called a "vicious cycle"—the difficulty of persuading distributors and retailers to handle programming as long as the machine population remained so sparse and the difficulty of expanding the machine population as long as the availability of programming was suspect.

Besides, the list of possible white knights was thick with companies that had already passed up the chance to become involved with Cartrivision. It had been Stanton's original dream to persuade all the big names in the TV business—RCA, Zenith, GE, Philco, and the rest—to incorporate his system in their sets and manufacture their own recorder-player decks. Stanton and Co. expected to make their money marketing the software, not the hardware. (Avco, conveniently, owned a movie company, Avco-Embassy.) "The point I took was that America's lead was in the software," Stanton explained. "American software dominates the world. So whatever hardware system we all got together on here, the logic was inevitable that the Japanese would have to adopt us. The company dominating the software would be in a position to spread the hardware. We said, 'We'll give you a free license to make this machine.' Once the American TV companies agreed, the world would have to have followed. But wherever we went we ran into people who said, 'There's no future in it' or 'We can buy it from abroad.' And there was the NIH syndrome: not invented here. And the net result is that today there is not one home video machine made in America."

In addition to his other troubles, Stanton had to contend with a burst of announcements about a new home VTR that RCA, apparently cured of its infatuation with holograms and lasers, was getting ready to introduce "at an early date." To judge by the statements of RCA's marketing people, it would be a tough contender. Like the U-matic, it would use three-quarter-inch tape in a cassette with an hour's recording capacity, and word had it that a new, thinner tape would soon allow for two-hour recording. To top it off, the RCA machine would be sold at a price "significantly below that of any other video recorder-player introduced to date." The figure of seven hundred dollars was mentioned.

The Cartrivision people, well steeped as they were in the economics of VTR production, found it hard to understand how RCA could offer so much for so little, and they were troubled by a curious symmetry in the timing of RCA's promotional activities and their own desperate efforts to keep faith alive in Cartrivision. "It's tough to fight a ghost," Hilford told a reporter that summer as he and Stanton carried on their efforts to raise more money. "I don't believe it will come to market." As the months passed, the introduction date for SelectaVision MagTape, as RCA called this latest system, kept sliding back in time, and the scale of its planned introduction kept diminishing, until in October—when Cartrivision's demise seemed fairly irrevocable—RCA announced that it would settle for a "field test" early in 1974. Two hundred machines would be given to families in Indianapolis, and their reactions would be surveyed. On the basis of their reactions, RCA sent the project back to the drawing boards, and there it remained.

WHAT YOU WANT TO WATCH—WHEN YOU WANT TO WATCH IT

RESEARCH AND DEVELOPMENT is a jerky business. The inventor moves forward a distance and pauses to show his superiors what he has done. If they approve, he pauses some more, putting aside his latest thoughts and plans, and sometimes his doubts, in order to cooperate in the process of translating a prototype into a reasonable product. To a free-floating mind like Nobutoshi Kihara's, these interruptions were difficult. Every hour devoted to give-and-take with the manufacturing people and the marketing people was an hour not devoted to dreaming up more advanced products or major improvements in the product as hand. When other companies got into the act—as they did with Sony's decision to make the U-matic a three-way collaboration—the complications increased geometrically. Like his superiors, Kihara wasn't completely sold, at first, on the benefits of standardization. He was proud of the standards that Sony had set for itself and skeptical of the ability of a joint committee to live up to them. As the negotiations with Matsushita Electric and JVC stretched out, and as the U-matic gradually got altered and compromised in small but significant ways—the cassette growing to twenty percent more than its original size, for example, and the machine inevitably growing larger as well—he winced, registered his objections, and did his best to think about other things.

Many of the engineers who had helped Kihara develop the U-matic were gone now, detailed to service on the U-matic production line under Masahiko Morizono. "From product to product, my department has been something like a school," Kihara said later. "People come to my school and study new products for one, two, or three years, and then, when the product is finished, they go out into the production side—ten, twenty people at a time."

Freed of the burden of worrying about the U-matic and given a new student body to work with, he found his thoughts rushing forward. He tried to look on the U-matic with a suspicious eye—as a prototype to be criticized and rethought. He asked himself how the machine and the cassette could be made still smaller and less expensive, and his attention turned to an idea he had been forced to shelve two years earlier.

In a U-matic recording the video tracks are separated from each other by a space known as a guard band. Its purpose is to prevent cross talk, or interference, between one track and the next. To Kihara the guard band was like vacant land in the eye of a hungry real estate developer. The waste was all the more vexing because of the existence of a technique known as azimuth recording, designed to make guard bands unnecessary. In a two-head recorder—and most of the VTRs that had come along since the early sixties had been of the two-head, helical-scanning type—one head is responsible for the even-numbered tracks and the other for the odd-numbered. In azimuth recording the two heads are mounted at angles slightly off perpendicular to the tape, one slanted to the left and the other to the right. The result is a modest but tolerable reduction in the ability of each head, on playback, to pick up the track it is *meant* to pick up and a much more significant falloff in its tendency to pick up the adjacent track—the one laid down by the other head. Azimuth recording is the electronic equivalent of putting blinders on a racehorse so that it can compete in tight formation, keeping its eyes resolutely forward.

The concept had been patented in 1958 by Shiro Okamura, a professor at the Tokyo University school of communications. Matsushita

Electric had marketed an azimuth-recording VTR in 1968—a two-hour black-and-white model—and at about the same time Kihara had tried to use the technique in the machine that later became the U-matic. The result had been what he recalled as a "big noise" in the color, forcing him to revert to the customary guard bands. Azimuth recording does not work so neatly with the color portion of the video signal because much lower frequencies are involved, and the loss of response on which azimuth recording depends varies with the frequency. But Kihara had always regarded the problem as solvable, and now that the U-matic was behind him, he began to think about juggling the phase of the color signal as a possible solution. When two parallel tracks of a magnetic recording are "in phase," they match up perfectly, like pieces of lasagna with the ridges of one nestled inside the grooves of the other. Putting them "out of phase" is like shifting the position of one piece slightly; it creates a separation. In theory, it seemed to Kihara, such a separation could be used to eliminate the problem of color cross talk. He set out to turn the theory into practice.

The three-way agreement on U-matic left its signatories free to develop new formats, and the top people at Sony were quick to realize that this was not the consumer product they had been striving for. Soon after the U-matic had gone on the market, Masaru Ibuka returned from a trip to the United States carrying a paperback book. The U-matic cassette was hardcover size. A true consumer videocassette ought to be paperback size, Ibuka declared. Now Kihara had a mandate. Many elements would have to fall in place to make such a small cassette possible—not only the reconciliation of azimuth recording and color but advances in tape making, integrated circuits, servo mechanisms, and head and drum construction. Sony's next generation of video machines would be the product of a long, collaborative effort guided by a sure sense of the objective and the patience to wait until each specification had been met.

These qualities set Sony apart. Few of its competitors knew the market or the technology as well as Sony did, or were able to define the goal as clearly. The American efforts in particular had tended to originate in the laboratories and executive suites of companies with next to no history of actually making and selling VTRs, and they were conceived on such an expensive scale as to leave no room for trial and error in their sponsors' calculations—no opportunity to bring a flawed product to market (as Sony had done with the CV-2000), learn from it, and persist.

The efforts of some of Sony's Japanese and European competitors, on the other hand, suffered from an excess of trial and error. Matsushita had hardly begun to market its version of the U-matic when it came out with a half-inch cartridge-format color VTR, known as AutoVision. It was small and affordable—just what the U-matic wasn't. But the cartridges had a thirty-minute recording capacity, a limitation that "received severe criticism from the marketplace," as Hiroshi Sugaya, the general manager of Matsushita's planning and engineering division, said later. The Dutch company Philips, hoping to lead the way in videocassettes as it had in audiocassettes, put a one-hour machine on the market in 1972. The tape was a half-inch wide instead of three-quarters; but the recorder was not appreciably smaller or less expensive than a U-matic, and while it enjoyed some success in Europe, it had none in the United States. Two years later Toshiba and Sanyo joined forces to launch a half-inch system called the V-Cord, which found even less favor. It was intended for the institutional market, but the decision to use the skip-field technique of Cartrivision and the CV models made the V-Cord a far cry from the U-matic in terms of picture quality.

The industry seemed to be generating a new video format every few months, each saddled with its own particular failings in addition to the common failing of incompatibility. And the will to persist—*gaman*—was not always there. Most of the companies in the competition had never really made any money from VTRs, so their managers tended to take a suspicious view of VTR research. JVC, for one, had made significant technological contributions in the areas of helical scanning and color recording, but they were isolated developments, which the company had been unable to incorporate into a successful product. Shizuo Takano, the director of JVC's video products division, established a three-man team in 1971 to develop a true home VCR, and over the next two years he was forever pleading with his superiors merely for permission to continue the effort on that modest scale. Only at Sony did the top management look on the video division as a dependable source of profits. When it came to VTRs, the talk in the Japanese electronics industry (to quote Nakagawa) was: "Number one is Sony. There is no number two or three. Four is JVC. Five is Matsushita. And there is no six or seven."

By the middle of 1974 Kihara had built a prototype which fulfilled the requirements laid down by Ibuka. It was dubbed the Betamax, *beta* being a Japanese painting and calligraphy term for a brushstroke so rich that it completely covers the surface below, just as the video signal would now completely cover the tape. In some ways the Betamax was a scaled-down U-matic. By coming up with a gentler and more exact tape transport and guide system and a new formula for the binder, which reduced the tendency of the tape to stick to the head, Kihara had managed to decrease the thickness of the tape by twenty-five percent. He had cut the track width from eighty-five microns to less than sixty, largely by incorporating a smaller recording head and using tape with magnetic powder made from a cobalt alloy. These refinements, along with the key breakthrough of azimuth recording, had paved the way to an overall drop in the rate of tape consumption of about seventy-five percent, from 2.81 square inches to 0.78 square inch a second. (The three-quarter-inch tape of the U-matic moved at a speed of 3.75 inches a second; the half-inch tape of the Betamax at 1.57 inches a second.) The Betamax cassette, with a recording capacity of one hour, was indeed the size of a paperback—scarcely more than a third the volume of a U-matic cassette—and it could be manufactured for half the price. The machine, when it went into production, would weigh forty pounds instead of the U-Matic's sixty. Equally important from a cost standpoint. Kihara and his colleagues had integrated the circuitry into as few units as possible, drawing on a body of expertise that Sony had developed while making its well-regarded Trinitron TVs.

Everyone who counted at Sony now believed that the long-cherished goal of a VTR for the home had been realized. But for the Betamax to fulfill this high hope, Ibuka and Morita thought that it needed one more attribute: compatibility. They wanted to make the Betamax—

like the U-matic in its corner of the market—an industry standard. And so, on September 29, 1974, the two vice-presidents of Sony, Kazuo Iwama and Norio Ohga, took a one-page diagram and a sample cassette to Matsushita Electric's headquarters in Osaka and proposed that the two companies (along with Matsushita's affiliate JVC) jointly adopt it. Matsushita was three times as big than Sony. With Sony's technological skills and Matsushita's marketing clout, the Sony people reasoned, the Betamax would be an unbeatable contender against any home videocassette recorder (or the equivalent) that another company—American, European, or Japanese—could possibly develop. A week later Iwama made the same presentation to the top people at JVC. Sony intended to put its first Betamax models into production by the end of the year, but the company took the position that Matsushita and JVC would not suffer by coming along a few months later because history showed that the demand for a new product built slowly. All three companies, Sony argued, would be starting off on essentially the same footing.

More than a month passed, however, without any response from either Matsushita or JVC, and in the middle of November, impatient for a resolution, Morita went to Osaka to talk president-to-president with Masaharu Matsushita of Matsushita Electric. "I told Mr. Matsushita that the home videocassette recorder will be the biggest item next to the color TV in the whole next generation of home entertainment," Morita recalled. Word came back that the engineers at Matsushita Electric had been unable to comprehend the details of Sony's design merely from a diagram and a single cassette. The company would need to see an actual machine before it could decide. On December 4 delegations from Matsushita and JVC were received at Sony headquarters in Tokyo and treated to a demonstration. Ibuka and Morita had been nervous about disclosing so much before they had a commitment. But they took solace in their faith that a Betamax in action would surely resolve whatever doubts the people at Matsushita and JVC were feeling.

It didn't. The JVC people responded perfunctorily, as if they weren't terribly interested. The Matsushita people, at least, raised some questions. Sugaya, the general manager of Matsushita's planning and engineering division, brought up the one hour recording capacity. Matsushita's market surveys, he said, suggested that a home video machine ought to be able to record a two-hour program, and he asked if Sony would consider altering its format to that end. He was told that Sony had examined the market and had determined that an hour would be enough—at least to begin with. Sony took the position that a small and inexpensive cassette was a higher priority than recording time. In any case, the company had already cast dies and made other preparations to begin production, so it was too late in the day for fundamental changes.

Once again Sony received no clear response to its offer—"We're still evaluating it," the Matsushita people kept saying—and the management began to get fidgety. If they delayed much longer, Ibuka and Morita were afraid the Betamax would lose its competitive edge. Sony prided itself on being ahead of other companies in product development, but what was the point, they had to ask themselves, if they didn't market a product when it was ready for the market? Sony was lucky that none of the other attempts at a consumer VTR had caught on. There would be more where they had come from.

After six months of waiting, Sony's patience ran out. On April 16, 1975, Morita held a press conference to announce the introduction of the first Betamax—the SL-7300 deck. Kihara was on hand to demonstrate what a Betamax could do, and an observant American reporter asked if he was, in fact, the inventor of the product. Morita gave the reporter a lecture in reply. It would be bad for company morale, he said, to single out an individual engineer's part in a group effort.

The SL-7300 cost 229,800 yen (when U-matics were selling for nearly 400,000 yen), and Japanese consumers found it much more pleasing than any of the previous "home" VTRs which had come their way. The Betamax might not be for everybody—not yet—but railroad engineers, airline pilots, and other high-salaried types with odd working hours took a strong interest in the ability to turn anytime into prime time.

In the United States the marketing began with the console model SL-6300, which came with a nineteen-inch Trinitron included. At $2,295 the SL-6300's appeal was limited. The real test, Sony knew, would come with the arrival of the Betamax deck, and when it went on sale in February 1976, at $1,295—a price that, with the severe inflation of the early seventies, was comparable to what black-and-white and color TVs had cost in their infancy—interest increased dramatically. "It was fantastic, really," Harvey Schein of Sonam said later. "When you have a new product that is as jazzy as a videotape recorder, you really skim off the cream of the consuming public. The Betamax was selling for over a thousand dollars, and the blank tapes were fifteen or sixteen dollars each. But there were so many wealthy people who wanted to be the first in the neighborhood that it just went whoof—like a vacuum. It flew off the shelf, and then the tapes flew off the shelf as well, and there was a tremendous shortage of tape. At one point, I remember, we were amazed that we had sold something like twenty-five thousand videotape decks, and we had about a quarter of a million tapes to support that. We had expected people to buy six or eight blank tapes with each machine. Instead, they were buying twelve or fifteen or more. And lo and behold, all the tapes were all gone! We got letters and telephone calls from consumer protection agencies around the country saying, 'How dare you come out with a razor without having enough razor blades?' Of course, like all such things, it began to taper off after a few months."

When it did, Sony launched a nationwide ad campaign. The opportunity to promote a completely new product is a rare and marvelous event in the life of an advertising man, and Michael Mangano, the principal writer on the Sony account at Doyle Dane Bernbach, had a high time of it. The Betamax was the "most revolutionary product I've ever worked on," he said later. Simply explaining what it did was a challenge. Customers who came across it in stores tended to ask questions like "Where do I get the picture *developed?*" They had to be told that videotape could be played back instantly and that a Betamax, with its optional timer, could be set to record a program unattended. They had to have some glimmer of an understanding of the notion that a TV consisted of two major elements—a monitor and a tuner—and the fact that a Betamax came with its own tuner built in.

Mangano and his associates tried to boil it all down to essentials and get a few laughs in the process. In addition to print ads like the one that provoked Universal to sue, they conceived a series of memorable TV commercials. In one, a New York City cabdriver, getting off work at dawn or so, proudly announced, "I'm going home to watch 'The Late Show,' " and the camera panned across a vista of amazed expressions on his co-workers' faces. In another, Count Dracula, also returning home late, flicked on his Betamax and explained in a Transylvanian vibrato: "If you work nights the way *I* do, you miss a lot of great TV shows. But I don't miss them anymore, thanks to Sony's Betamax deck which hooks up to any TV set. While I am out, Betamax is automatically videotaping my favorite show for me to play back when I get home. And now I'm going to watch it." The theme from "Alfred Hitchcock Presents" poured forth as the commercial ended.

In a third commercial a compulsive sports fan who had yet to acquire a Betamax switched nervously back and forth between a basketball game and a baseball game while a narrator soberly inquired: "Ever want to watch two things that are on at the same time? Well, now you can. Because Sony's revolutionary new Betamax deck, which hooks up to any TV set, can actually videotape something off one channel, while you're watching another channel. With Sony's Betamax, you won't miss a thing." As the narrator finished his explanation, the sports fan could be seen switching back to the baseball game again, just in time to hear the commentator announce, "A triple play! You've just seen a triple play! You don't see *those* very often."

"By January or February of 1977," Schein recalled, "one of the trade magazines in the hardware field did a survey, and would you believe it? Forty-nine percent of all the people in the United States knew what a Betamax was." They knew what it was partly because, unlike Cartrivision, it was advertised for a single, definable purpose—namely, recording programs off a TV set and playing them back later. Akio Morita neatly summed up the Betamax's mission with the phrase *time shift,* which he coined at a luncheon with executives of Time Inc. People might find certain things about the Betamax hard to understand, but they understood the basic idea, and they liked it. Sony had tapped into a desire which had been gathering force, in a stifled and largely unconscious way, all through the thirty-year history of television.

It seemed to Harvey Schein, however, that Sidney Sheinberg and his colleagues at Universal deserved some of the credit. "One of the most important elements in the success of the Betamax, I believe, was the lawsuit that Universal brought," Schein said. "The publicity that we got as a result of that was absolutely phenomenal, and it had a tremendous psychological effect. I debated with Sid Sheinberg on 'Walter Cronkite.' He called me a 'highwayman.' He equated Sony with the highwaymen of old who stole people's property, and I said, 'Wait a minute. What right do you have to tell me or anybody else what I can or cannot do in the privacy of my bedroom? [When the defenders of home taping got to hypothesizing, they tended to put the VCR in the bedroom, although studies would show that about 85 percent of VCR owners kept their machines in the living room or the family room.] Are you trying to tell me that the next time I watch a Universal program I've got to sit with my hands folded? Are you telling me . . . if I'm watching "Columbo" and I've invested fifty-five minutes in it, and he's just about to reveal who the murderer is, and I'm called to the telephone—or nature calls—that I can't flip on my tape recorder? Would you want me to watch the whole thing all over again to see what happened—six months or two years later, or whenever it comes up on reruns?' And he said, 'Yes.' I said, 'You must be kidding!'

"When people saw how worked up the movie industry had gotten," Schein went on, "they said, 'Wait a minute! These guys here must be offering us something worthwhile if those people out in Hollywood are so afraid.' And the other perception was: 'Well, Jesus, we'd better go out and buy one soon, because otherwise we're not going to get the chance.' "

A TRIAL LAWYER'S DREAM

WHEN HARVEY SCHEIN set out to recruit a team of lawyers to defend Sony against Universal, he didn't have to look very far. Edward Rosiny, of the New York law firm of Rosenman, Colin, Freund, Lewis, and Cohen, had been Sony's principal legal representative in the United States since 1960, when he had won his way into Akio Morita's heart by buying the company out of a bad contract for seventy-five thousand dollars instead of the two hundred thousand Morita had resigned himself to paying. Rosiny had severe emphysema and was getting on in years, and copyright was not his specialty. But the firm had a resident copyright expert, Ambrose Doskow, who had been consulted by Sonam in the past—on the legality of home taping, among other issues. So it was a matter of course that Rosenman, Colin, Freund, Lewis, and Cohen would handle the defense, with Rosiny playing a senior statesman role and Doskow in direct charge. Rosiny had joined Rosenman, Colin, Freund, Lewis, and Cohen in 1972, bringing Sony's business and his own talents to the firm as a package deal. If his partners had never had any previous reason to congratulate themselves on the wisdom of taking Rosiny into the fold, they had one now; whatever the outcome of *Universal* v. *Sony,* it seemed safe to predict that Rosenman, Colin, Freund, Lewis and Cohen—like

Universal's law firm, Rosenfeld, Meyer, and Susman—would come out ahead on the deal.

But Sony would also need a Los Angeles lawyer—a "local counsel"—and for that role Rosiny recommended his old friend Dean Dunlavey. Dunlavey had represented Sony on several matters, including the Ampex litigation, and Rosiny knew that Morita held him in high regard. Ira Gomberg, Sonam's general counsel, had also formed a favorable impression. What stuck in his memory was Dunlavey's willingness, when consulted on a fine point of law, to admit that it was outside his experience and, if need be, to recommend another lawyer—maybe even one with another firm. "Most lawyers today are interested in pushing their own firms," Gomberg explained. "They're 'experts' in everything, or they're willing to be trained at your expense."

When Dean Dunlavey got involved in the Betamax case, he was fifty-one and a partner in one of the country's most respected law firms, Gibson, Dunn, and Crutcher, from whose ranks Ronald Reagan would recruit an attorney general (William French Smith) a few years later. Dunlavey had a secretary and a spacious office on the forty-fifth floor of the Crocker Center, a brown-and-green slab of a building in downtown Los Angeles. He was a graduate of Harvard University and the Law School of the University of California at Berkeley. He had argued two cases before the United States Supreme Court. He was everything, in short, that a corporate client could expect of a lawyer—until the client got a look at him.

A large man, built on lines unrecognized by the fashion industry, Dunlavey seemed uncomfortable in a suit and tie, and the suit and tie seemed equally uncomfortable. In thought, he would shut his eyes and take on a strong resemblance to a hibernating bear. In his active periods, which were the norm, the eyes were rarely more than halfway open, so he generally gave the impression that he had just got out of bed or had just emerged from his cave after a long winter. "He's a rough character who happens to find himself in a very elite law firm," a Washington lawyer who crossed paths with Dunlavey observed. "You meet him and he looks like a guy who drives a pickup truck."

At Gibson, Dunn, and Crutcher, Dunlavey was "kind of an anomaly," by his own description. "I'm a trial lawyer who still likes to go to court in a firm of corporate lawyers who like to settle out of court," he explained. "*Universal* versus *Sony* was a trial lawyer's dream, because Universal could not speak for all copyright owners and Sony could not speak for all manufacturers. So there was no way they could settle this case."

If the young Dunlavey (and it wasn't easy to imagine him young) had ever been afflicted with a Harvard accent, it had been purged from his system by what he called the "formative experience of my life"—his service as a combat officer in the Pacific during the Second World War. In the Philippines he got to know a number of Japanese soldiers as POWs. Once he told a prisoner to pump some water, forgot about it, and found the man lying next to the pump, unconscious, five or six hours later. "He'd cranked that thing until he'd passed out," Dunlavey recalled. "They are the most dedicated and disciplined people I've ever seen." The army was the alma mater to which Dunlavey remained most loyal. During the Vietnam War years he paid a visit to Cambridge, Massachusetts, and happened on an antiwar demonstration. He felt the urge to lead his old army unit through Harvard Square "and mow them down."

Dunlavey majored in chemistry at Harvard, received a Ph.D. in nuclear chemistry at Berkeley, and briefly worked with Glenn Seaborg at the Berkeley Radiation Laboratory. But in 1952 he belatedly enrolled as a first-year law student. Science was too calm and collegial, its rewards too far-off and uncertain, Dunlavey had decided. Law was more combative and "it paid pretty well and you could be your own boss." He was disappointed in his fellow Berkeley law students. "They were certainly not hard-core thinkers like I was used to in chemistry," he said, "but these people controlled society."

Lawyers, like scientists, have become more team-spirited in recent years—especially when involved in cases that call for a lot of research and deposition gathering. Dunlavey is a throwback. "He comes from the era of the rough-and-woolly trial lawyer who goes in and solos it all the way," a junior colleague at Gibson, Dunn, and Crutcher said. Dunlavey himself explained his position thusly: "If you're going to go to trial, you can't use a lot of lawyers and paralegals to do your preparation for you, because when the time comes at trial that you're going to need it, you haven't got it. It's in somebody else's head."

During the first few months of Sony's preparations a belief took hold in certain quarters—the quarters of Rosenman, Colin, Freund, Lewis, and Cohen in New York City, to be exact—that Dunlavey would be a kind of colonial satrap taking his orders from the imperial headquarters back East. This belief subsided in the spring of 1977, when Akio Morita, on one of his regular transpacific excursions, received a briefing on the Betamax case from Ambrose Doskow. The lawsuit was a complicated affair, Morita learned, or perhaps it only seemed complicated when Doskow explained it. Slight and scholarly, well steeped in the history of copyright, Doskow had represented the Metropolitan Opera in a landmark suit against a defendant who sold unauthorized tapes of the Met's radio broadcasts, and he felt a need to escort Morita through the precedents, the doctrines, the uncertainties, and the contingencies. He was an "on the one hand—on the other hand" type whose nickname, Amby, could hardly have been more apt. Morita listened to everything Doskow had to say, and then, with the machetelike simplicity that only a chief executive can bring to a problem, he cut through to a conclusion. He would put his faith in Dean Dunlavey henceforth.

Doskow receded into the shadows. "Mr. Morita felt he wasn't sufficiently jugular," Schein explained. Harvey Schein's star also went into a tailspin around this time. Schein had been given a good deal of autonomy, and Morita did not altogether like the way he had used it. "The trouble with American management of the Sonam operation in the early days under Schein was that profit was the main goal," Morita wrote in his book *Made in Japan*. "In my view, profit doesn't have to be so high, because in Japanese companies our shareholders do not clamor for immediate returns; rather they prefer long-term growth and appreciation. . . ." Morita and Schein had had a number of quarrels on this theme, including one over the amount of money to be spent on Betamax advertising. That particular dispute had ended with Morita's yelling, "If you're not going to spend a million or two million dollars on the Betamax campaign in the next two months, I will fire you." Schein had spent the money, but Morita later discovered that some of it had come out of the advertising budget for other products. Early in 1978 Schein ejected from Sony and, with a modest golden parachute to cushion his fall, landed in an executive vice-presidency with Warner Communications.

"Just simply, I trust him" was how Morita explained Dunlavey's selection as the lawyer to defend Sony against Universal. The Americans at Sonam were used to whirlwind visits by Morita which sometimes resulted in abrupt and seemingly arbitrary decisions. So the disappearance of Doskow, the ascent of Dunlavey, and the decline of Schein fell into a familiar category of developments which set shoulders to shrugging and heads to shaking, with unspoken reference to bad karma and the mysterious wisdom of the East. But when all these adjustments had been accomplished, they proved to have a clear and perhaps not altogether irrational effect on the Betamax litigation: Sony had exchanged a specialist for a generalist, a theoretician for a scrapper.

"I've tried personal injuries, malpractice, patents, and the gamut," Dunlavey said. "I don't think there's a civil lawsuit of any kind that I haven't tried, and I'm about to beat a drum that I beat whenever I have the chance. If you want advice in an area of law to govern your conduct, then you would normally and properly go to a specialist in that field of law. If you want to know, 'What are the consequences if I do this or that?' about a copyright issue, go to a copyright lawyer. But once you're in a lawsuit, it doesn't make any difference what the subject matter of the lawsuit is. You've got to have a man who knows the judicial process and knows how to best get what you want. He'll learn the law that pertains to your case in a relatively short time."

As a nonspecialist Dunlavey came to the lawsuit with a pronounced lack of respect for the intellectual rigor of copyright law. To his way of thinking, fair use wasn't a "doctrine." It was simply a phrase that had entered the legal vocabulary after countless occasions when judges had said to themselves, "Yeah, this is copyright infringement by any normal test. It was copyrighted. It was copied. But it just isn't fair to hold the guy responsible for infringement, and I'm not going to do it. So I'll call it fair use." If the phrase had never yet been applied to videotaping TV programs in the home, Sony's job was to find an analogy close enough to persuade a court of law that such a ruling would be no great departure from tradition.

The usual place to look for such an analogy would be in the case law—in court decisions touching on similar issues. In this instance, however, the strongest analogy was one that had never been tested in court. The invention of the audiocassette recorder in the early sixties, and the subsequent emergence of inexpensive, high-quality audiocassette recorders, had made music tapers out of millions of American teenagers and a fair number of adults. While the law didn't specifically sanction this activity—and the record companies weren't happy about it—it had never been seriously challenged, and there were signs that Congress (or at least some members of Congress) viewed it as benign. In 1971, during a floor debate on a record piracy bill which was later folded into the 1976 copyright statute, Abraham Kazen, Jr., a Texas Democrat, asked if he was "correct in assuming that the bill protects copyrighted material that is duplicated for commercial purposes only."

"Yes," answered Robert Kastenmeier, the Wisconsin Democrat who chaired the subcommittee with jurisdiction over copyright.

"In other words," Kazen said, "if your child were to record off of a program that comes through the air on the radio or television"—he meant the sound track of a TV program, not the picture—"and then used it for her own personal pleasure, for listening pleasure, this use would not be included under the penalties of this bill?"

"This is not included in the bill," Kastenmeier replied. "I am glad the gentleman raises the point. . . . This is considered both presently and under the proposed law to be fair use."

When the law is unclear on its face, the courts look to what they call the legislative history for evidence of Congress's intent. Behind every ambiguity, it is assumed, lies a clear congressional purpose—something of a more substantive nature than, say, the desire to adjourn by Thanksgiving. And it is further assumed that if enough lawyers spend enough time digging through hearing transcripts, committee reports, and discarded drafts of bills, the intent of Congress will stand revealed for all to see. An exchange between two members of the House in 1971 might strike the lay observer as slight evidence of what Congress as a whole meant to do when it passed a law five years later, but in the world of legislative history the deepest finds are often the most prized.

Mucking through old congressional documents was one assignment Dunlavey was willing to delegate. "It's a dull, thankless job," he said, adding, as if it were the tail end of a syllogism, "We had a young kid from Harvard who did it for us."

But the plaintiffs had well-educated kids of their own, who were equally busy unearthing contrary evidence. Congress, the lawyers for Universal and Disney concluded, had gone out of its way to distinguish the treatment of audiovisual and audio recordings. Audiovisual works had specifically been excluded, for example, from a provision of the law allowing broadcasters and transmitters to make emergency backup copies, known as *ephemeral recordings.* In a 1967 report of the House Judiciary Committee, Universal's probers found a statement about "the special nature of motion pictures and audiovisual works," which made them "exceptionally vulnerable to copyright impairment under an ephemeral recording exemption. . . ." The plaintiffs would also emphasize that the law gave the creator of an audiovisual work authority over performances as well as duplication and distribution. This "performance right" was another form of protection not attached to to records and audiotapes (or, in the law's parlance, "sound recordings").

As Universal's lawyers had anticipated, Sony's lawyers intended to put some legal distance between the company's role as the Betamax's manufacturer and the product's use to tape copyrighted programs off the air. There was much discussion of this issue—the so-called contributory infringement issue—during the early brainstorming sessions in New York. What right, Sony's lawyers asked, did Universal and Disney have to banish a product merely because it *could* be used to copy their works? A Betamax could also be used for making home movies or for recording programs that, for one reason or another, weren't protected by copyright. Somebody on the defense team raised the question, Would every copyright owner, in fact, object to having his work taped? "We believed there was a universe out there of people who were glad to have their programs copied," Ira Gomberg recalled. "Then we thought, 'Well, who would those folks be?' "

The world of nonprofit television suggested itself, and Ambrose

Doskow—who was still on the case at this stage—thought he knew of a prospect: Fred Rogers, the executive producer, head writer, and star of "Mister Rogers' Neighborhood," a popular children's series on Public TV. Doskow had represented Rogers in negotiations with the Corporation for Public Broadcasting. "I knew he was interested in getting as large an audience as possible," Doskow said later. "I knew what his position would be." A phone call confirmed it. The Betamax, according to Rogers's deposition, performed "a real service" by giving parents the flexibility to show programs to their children at "appropriate times."

It occurred to one of Sony's lawyers that since the shelf life of a football game is a brief one, the National Football League might approve of a technology that would allow a fan to see a game in spite of a conflicting engagement. Lawyers for the various sports leagues were approached, and within a few months the defense had procured a deposition from the general counsel to the National Hockey League, who testified that the NHL was indeed willing to have Betamax owners record televised hockey games for later viewing. More such depositions followed.

In the plaintiffs' camp Stephen Kroft and his associate, John Davis, a former Australian Olympic swimmer, were amassing evidence to show that the great majority of Betamax owners were recording programs whose copyright holders would and did object strenuously. Early in their search, they alighted on *The Videophile,* a cozy little periodical with a classified-ad section that had become a trysting place for video buffs out to get or give help in building collections of movies and TV shows. "Have lots to trade and access to cable TV," one advertiser boasted. Another asked for "any Garbo, Doris Day, Laurel & Hardy, Chaplin or Olivia Newton-John movies," *Psycho, Citizen Kane, M, The Seventh Seal,* episodes of "The Twilight Zone" and, above all, "anything on the Nuclear Power Plant accident in Pennsylvania."

The Videophile claimed to be "supported, in part, by a grant from the Land of Cotton Center for the Preservation of Popular Culture (old times there are not forgotten)," but it was basically a one-man venture, the one man being James Lowe of Tallahassee, Florida. A lawyer in his spare time, Lowe had taken the precaution of reprinting, alongside his classifieds, a warning usually found on motion picture prints: "Federal law provides severe civil and criminal penalties for the unauthorized reproduction, distribution, or exhibition of copyrighted motion pictures. . . . The Federal Bureau of Investigation investigates allegations of criminal copyright infringement." Next to this caution loomed, menacingly, the FBI logo. In his deposition Lowe confessed to possession of one video recording of a Universal property, *Psycho,* but vowed to "erase or trade" it as soon as he had a group of friends over to watch it.

Through *The Videophile,* Kroft also found Marc Wielage, a video equipment salesman who owned three Betamaxes and had a collection of three hundred tapes, including *The Mummy's Hand,* a Universal title, and Disney's *This Is Your Life Donald Duck.* Wielage had a Betamax-size padded suitcase, and he was in the habit of traveling from city to city and checking into hotels equipped with cable TV so as to gather new material for his collection.

In search of harder data, the plaintiffs secured a list of Betamax purchasers in the Los Angeles area and began to make a systematic study of their habits. But they had not gone very far with this effort when Dunlavey lodged a protest. "They had sent out a private investigator and some paralegals, and they were badgering homeowners," he said later. "So we went to the judge and said, 'Make them cut this out, this going from door to door, knocking and scaring people.' The judge said, 'Yes, that's no way to behave. If you want to find out what people are doing, I will let each of you conduct a survey, using a proper foundation, and you can then put those surveys in evidence.' " This was an invitation that neither side could resist.

The Betamax had, by now, graced the language with two new verbs—to *time-shift* (coined by Morita) and to *library* (authorship unknown). The defendants took the view that the typical user was a mere time-shifter, doing what the producers and networks meant him to do: watch a program once, albeit a few hours or days after it had been broadcast. The plaintiffs claimed that people were using their Betamaxes to build collections of copyrighted material and selling, trading, and copying each other's tapes. Universal engaged a prominent California polling firm, Field Research, to do its survey. In the summer of 1978 Field

Research conducted telephone interviews with "805 adults self designated as most familiar with the use of the video tape recorder in the household." The results showed that the average household harbored 31.5 cassettes—far more, Kroft could argue, than anyone needed for time-shifting purposes—while 18.99 of those cassettes were to be "kept in your library." Twenty-three percent of the people surveyed reported that they used their Betamaxes for librarying purposes more often than for time-shifting; fifty-four percent answered the opposite; twenty percent said it was an equal proposition.

Sony went to Crossley Surveys of New York, which conducted a thousand telephone interviews with "the individual in your household who uses the Betamax most often." The Crossley study concluded that (among other salient points) ninety percent of the respondents had used their Betamaxes for time-shifting; fifty percent had used them to record professional sports; and seventy-five percent of recordings were viewed only once. The two surveys cost upwards of fifty thousand dollars each and added a few more months to the pretrial maneuverings. But a few months no longer seemed very important.

"Big case" had been almost the first words out of the judge's mouth at the first pretrial hearing. The bigness of it worried Sidney Sheinberg. He had wanted to apply for a preliminary injunction in order to nip the Betamax in the bud, but Kroft had advised against it. The quest for an injunction might spin off into a drawn-out legal battle unto itself, Kroft felt. Besides, trial judges frequently issued injunctions only to be reversed on appeal. It would be better, he argued, to get all the evidence on the record before a higher court passed judgment.

Fortunately they were dealing with a judge who had a low tolerance for delay. Judge Warren Ferguson of the United States District Court for the Central District of California—who would be deciding the case as well as overseeing the trial, since none of the parties had wanted to trust such a subtle and complicated issue to a jury—was a Democrat who had been appointed by Lyndon Johnson. He had started out as a fairly stern upholder of the law across the board, but he was thought to have become more of a skeptic about the wisdom of Big Government after the death of his son in Vietnam. Only a week before the Betamax case arrived on his docket, Ferguson had decided another major TV-related lawsuit by declaring an end to the "family hour," an arrangement intended to cut a swath of innocence through the network schedule between 8:00 and 9:00 P.M. According to Ferguson's opinion, the family hour wasn't the voluntary arrangement the networks claimed it to be, but rather, as the producer Norman Lear and his coplaintiffs had argued, the result of improper pressure from the Federal Communications Commission.

During the family hour trial Ferguson was amazed to learn that network censors had removed the word *crud* from an episode of the series "M*A*S*H." To him, the word meant "the green mossy stuff that forms around wells, and it was only distasteful to me," he said, "because as a kid, one of my chores was to clean it out of the slurry." He had done his slurry cleaning in and around Eureka, Nevada, and "Having been born and raised in Nevada," he explained in the pages of *Who's Who in America,* "I have adopted an old prospector's philosophy: 'Live today; look every man in the eye; and tell the rest of the world to go to hell.' "

At the first hearing Kroft put in a plea for accelerated treatment, citing his clients' concern that the problem would only grow worse with time. Judge Ferguson promised to do what he could on that score. "He bent over backward," Dunlavey said, "because he saw the public impact." But it was a big case, as the judge had observed, and it would take more than a few words of judicial voodoo to shrink it.

True to the pattern of most lawsuits pitting large and wealthy interests against each other, the process that soaked up the most time and money was discovery—an elaborate legal board game in which each side scores points by obtaining documents belonging to the other while frustrating the other's efforts to do the same. From Sony, Universal put in a general request for materials touching on the company's internal discussions of legal questions involving the Betamax. "I suspected, and it turned out to be true," Kroft said later, "that Sony's files would be full of evidence that they knew they had a copyright problem." From Universal and Disney, Dunlavey hoped to secure documents showing a kind of implied consent to Sony's plans. When the suit had been filed, Morita had indignantly recalled a series of VTR demonstrations, going back a decade, which had been attended by Universal and Disney officials among other movie industry people. "Whenever we had a new machine demonstration, we always sent an invitation to Wasserman and Sheinberg," Morita recalled. "After one demonstration, Mr. Wasserman wanted to see me, and we had a meeting. He showed great interest in selling movies on video. He never said, 'This is illegal.' " (Whether Sony would have done anything differently if Universal and Disney *had* objected was, of course, another matter. When the question was put to Morita in 1985, he flashed an expression of utter astonishment. "I have no idea!" he said. "Such an assumption!")

The discovery process strained nerves as well as time and money. "The foreplay that goes on is unconscionable," Dunlavey said later. "You go back to court time and time again, and finally you flush out of the Universal files these documents where some guy says, 'Guess what I've seen today—a machine that'll record off the air. It may be ten years away from the market, but boy, is it going to do this or do that.' " In 1965, Dunlavey found, Wasserman had been invited to attend a demonstration of the CV-2000. "It can record television programs for immediate playback or later viewing," the invitation announced (although there was no indication that Wasserman himself ever saw it). "You can imagine how easy it will be to build a home library of outstanding television programs. . . ." An MCA subordinate attended the demonstration and reported back: "Probably won't be sold as a mass consumer item for the next five years. Presently a prestige 'home' item." A Disney official attended another such affair and sent a report to Walt Disney, along with a covering letter: "Dear Walt, Thought you might be interested in the attached. I saw a demonstration and I was very much impressed." Sony found some of this material in its own files. Joel Sternman, a lawyer with Sony's New York law firm, made a trip to a warehouse in Queens along with a paralegal, and they searched through room after room for Sony's copies of the invitations and correspondence. "When we dug out these documents, we were amazed," Sternman said later, "because we had previously asked Universal and Disney if they had any such materials and we had pretty much come up empty."

As any practiced hand in the discovery game knows, there is a time for withholding documents, a time for turning them over, and a time for turning them over with a vengeance. One of Universal's requests ended in a room in Tokyo—a room "completely filled floor to ceiling with documents," Kroft recalled, "which I'm supposed to go through for the next day. And to make it worse, they were accounting documents and they were in Japanese. I just started laughing." When he had finished laughing, he hired a team of translators and "got them in and made them instant paralegals, and the bottom line was I ended up getting plenty of helpful information."

Universal wound up submitting more than a thousand separate documents to the court. Sony, for its part, had over fifty Bekins boxes. Between them, they were prepared to put 145 witnesses on the stand.

ALL PROBLEMS HAVE TO BE SOLVED

THE TRIAL OF *Universal* v. *Sony* got under way on Tuesday, January 30, 1979, with the opening statements of the two attorneys and then—kicking off the case for the plaintiffs—a demonstration of the powers of a Betamax. "Rarely has 'The Mickey Mouse Club' had such a high-priced audience of adults," *Daily Variety* reported. "More than a dozen attorneys (probably charging $100 to $200 an hour), a federal judge and assorted clerk aides watched intently yesterday as a Betamax poured forth the familiar strains of 'M-i-c-k-e-y M-o-u-s-e.' . . ." One lawyer squatted on the floor for a better view and received a lecture from the judge about courtroom demeanor. Then Donn B. Tatum, the chairman of Walt Disney Productions, took the stand as a character witness for Mickey, Donald, and the rest of the gang. They were still a feisty bunch at the box office—that was the essence of Tatum's testimony. He told the court that *Snow White and the Seven Dwarfs*, a forty-year-old movie, had just taken in another twelve million dollars in its latest rerelease. Tatum was so alarmed about home videotaping, he said, that he had refused to allow *Mary Poppins* and *The Jungle Book* to be shown on QUBE, a cable TV operation in Columbus, Ohio, because he had determined that its thousands of subscribers included "at least six, perhaps maybe twelve" owners of videocassette recorders. Passing up the QUBE deal and a similar offer from Home Box Office had cost the company more than two million dollars.

On cross-examination, Dunlavey asked if Tatum could point to any *other* losses sustained by Disney—any of an *involuntary* kind. Tatum could not. He was also subjected to some vigorous questioning by the judge. Was it right, Ferguson inquired, for the government to tell people how to watch TV programs in their own homes? Tatum launched into a response on the theme of balancing privacy rights against a creator's right to control his work. "Does that include the right to tell a viewer *when* he must see it?" the judge asked. Tatum tried to explain that moviemaking was an unusually complicated and fragile enterprise that could not be sustained without generous legal protection. After all, he said, retreating into what he must have regarded as uncontroversial territory, "There are more intangible elements involved in the making of films than there are in the typical manufacturing kind of business—things called talent."

"Well," Judge Ferguson said, "it takes just as much talent to get your shoes shined."

"It's a different kind of talent," Tatum answered diplomatically.

The next witness was Lew Wasserman, and he also had a story to tell. "It has been printed in the Hollywood trade press," Wasserman told the court, "that when 'Gone With the Wind' was put on television, there was not a blank cassette available for sale in the United States for video recorders. Now, if everyone in the United States ultimately has a copy of 'Gone With the Wind,' there would not be much value in 'Gone With the Wind' being on television." Wasserman segued into a professorial talk on the hit-and-miss economics of the movie business and the need to exploit every potential market, from initial theatrical release to TV syndication. With the Betamax in the picture, it was "only common sense" that much of the revenue that follows a movie's sale to network TV would be lost. "The damage could be enormous," he said. But Ferguson seemed to pay closer attention when, in response to a question from Dunlavey, Wasserman acknowledged that Universal's theatrical film division had just had its best year ever and when he volunteered that the history of show business was full of gloomy predictions that had failed to come true. "They forecast the doom of radio stations when television developed on the horizon," Wasserman said. "Radio stations are more profitable today than they have ever been."

In his opening statement Kroft had ticked off a list of points he hoped to prove—and one that, he said, he wouldn't have to. When copyright is violated, he told the court, the law, out of respect for the economic unpredictability of artistic endeavor, does not require the victim to show harm. "There is a presumption that irreparable harm will occur," he said, "and we believe that shifts the burden to the defendants. . . ." It was true that the fair use section of the new copyright act included a reference to "the effect of the use upon the potential market for or value of the copyrighted work." But Kroft took that to mean that an accused infringer, in order to establish fair use, had to show he *hadn't* caused harm, while satisfying the other fair use criteria as well. Home videotaping did not even come close to meeting these requirements, Kroft believed, and that being so, the simple fact that Betamaxes were used to duplicate copyrighted works, along with the equally simple fact that Sony built and sold them for that purpose, added up to a simple case of copyright infringement.

It was a strong presentation—orderly, assured, and smartly delivered. A stranger happening into the courtroom without prior knowledge of the law or the facts would have been impressed—not only by the words but by the well-appointed young man who spoke them. Unfortunately the stranger who counted was the one on the bench, and it quickly became apparent that harm, whatever its status in the law, occupied a prominent place in his thinking. As the trial proceeded, Kroft found himself devoting more and more time to making the point he felt no obligation to make. One expert witness after another tried to explain the complexities of the film and TV businesses and the threat posed by videocassette recorders to the "marketing sequence." And one expert witness after another was pummeled by the one-two punches of the attorney for the defense and the judge.

"We know beyond any peradventure of a doubt," said Jack Valenti, the president of the Motion Picture Association of America—and a witness who contributed some of the trial's lushest verbiage—"that people are buying large numbers of videocassettes; ergo, they are not just using them to record and erase." The Betamax was a "parasitical" device, he said. He agreed with Ferguson that the courts shouldn't "intrude in the home" but thought there should be some compensation. In West Germany the government had created a copyright royalty on every videocassette recorder sold, Valenti pointed out. Without some such arrangement, "the public in the final end may be the loser because programs do not come from the tooth fairy. They come from people who have risked money up front, hoping they can recoup it somewhere down that marketing-sequence line."

"You are aware, are you not," Dunlavey asked when his time came to question Valenti, "that every once in a while President Carter gives a half-hour address to the nation where he is talking about some topic of current interest?"

"Yes, sir," answered Valenti.

"He seems to have a penchant," Dunlavey said, "for starting these talks about nine o'clock in the East, which is six o'clock in the West, and I am on the freeway going home about that time—if I am lucky. Is there anything wrong, if I had a Betamax, in my recording his speech while I am on my way home and then viewing it after I get there?"

Valenti responded that "if it were possible to restrict" the use of Betamaxes to "the recording of presidential addresses, I personally would have no problem with that."

Judge Ferguson offered up another hypothetical. What if a worker on the night shift wanted to use a Betamax to record a program at eight o'clock in the evening and watch it at two in the morning? Where was the harm? Why should the copyright owner be compensated?

Valenti embarked on a wide-ranging answer about the "rostrum of free television" and its dependence on the ability to measure not just how *many* people watched a program but *when* they watched it. But Ferguson wanted to know—"Yes or no"— if the situation he had imagined was harmful. Valenti, rather ingeniously, responded that the night shift worker in such a case would be unable to see the regularly scheduled programs at 2:00 A.M., which would harm the producers and sponsors of those programs, who were "directly targeting to people who are working at night."

"There are always distractions," Ferguson observed.

"We are on the ragged edge of a vast philosophical enigma that really has to be decided," Valenti observed, "because it cannot be left out there naked and alone just to fester."

"All problems have to be solved," Ferguson remarked.

"I beg your pardon?" said Valenti

"All problems have to be solved," the judge repeated.

"I am willing to suggest that is true," Valenti replied.

Sidney Sheinberg took the stand on the twelfth day of the trial, and he, too, found himself in a debate with the judge. Testifying after a man from Field Research had delivered a statistical report on the amount of librarying and deleting of commercials, Sheinberg commented that the truth was undoubtedly far worse than any survey suggested.

Judge Ferguson asked if he meant to imply that people had *lied* to the surveyors.

Sheinberg said he would prefer to say that they "may not have volunteered the entire truth. . . . I am trying," he explained, "to put myself in the position of one of these people who receives one of these phone calls . . . and is asked, 'Do you have one of these? Do you have any movies in the library?' He may very well have heard of this case. For all he knows, somebody is liable to hold this against him at some point. I am just not sure that that kind of survey can be given the kind of credibility that asking for what kind of soap they use can be given. . . ." Betamax owners, Sheinberg said, had been "incentivized to zip right past the commercial," and even if they did watch the occasional commercial, there was no way for the networks to measure an audience that might see a program "any time from the following day to the following week to the following year to never."

"One never thought Sony would invent Betamax either, a few years ago, or that we would land on the moon," Ferguson said. "Don't you think that they have the mechanical ability and the intelligence to figure that problem out if they want to?"

Sheinberg didn't think so. "And to be even more candid, your honor," he said, "I do not see why the burden should be on the plaintiffs in this case to worry about that."

"Just because I asked a question do not get too defensive," Ferguson said.

"All right," Sheinberg said.

When it was Dunlavey's turn to question Sheinberg, he pointed out that videotape recorders had been widely used by people in the movie industry for years, and he asked if Sheinberg had ever used the Sony U-matic in his office to record a TV program. Sheinberg conceded that he had once recorded "a few minutes of something off-the-air to see if it worked." Dunlavey noted that Universal, once again, had two shows on the air at the same time—the miniseries "Centennial" and the series "Battlestar Galactica." "Would you not rather have somebody see both of those shows," he asked, "by recording one with a Betamax, and watching it at nine o'clock that same Sunday rather than missing it entirely?"

"No" was the long and the short of Sheinberg's reply. He was "as tough as a two-dollar steak," Dunlavey said later, making the phrase sound like a term of endearment.

On February 16, Paul Ruid, Universal's detective, recounted his visits to the various Betamax retailers, and Judge Ferguson announced that the coming Tuesday would be a day off in honor of the ninth anniversary of the death of his son in Vietnam. "If you have a spare moment or two and think about the young people of the world," he said to the lawyers, "wish them well."

William Griffiths, the nominal defendant, had prepared no defense and retained no counsel, and when called to testify, he readily confessed to having made a hundred recordings or so, including two of Universal's works—an episode of "Baa Baa Black Sheep" and the movie *Never Give an Inch.* Griffiths sailed through his testimony without a hostile question from anybody, but it must have been unsettling to the plaintiffs to hear their handpicked consumer testify that although he had bought his Betamax with the idea of building a library, he had settled for a career as a time-shifter because the tapes cost too much.

Three weeks into the trial Kroft introduced the first of a series of witnesses who would explain the economics of commercially supported TV and the danger posed by home taping. Harm might not be a necessary element of infringement in Kroft's mind, but a trial lawyer must argue in the alternative. First he makes the case for Argument A, and then he says, in effect, "And if you don't like Argument A, here's Argument B." Thomas Ryan, an advertising executive with the Gillette

Company, was prepared to testify to the various ways in which the Betamax's existence made it harder to sell razors. But Dunlavey lodged an objection. Any such speculation was irrelevant, he said, "because the Supreme Court has told us in TelePrompTer"—a cable TV case—"that in a copyright suit . . . there is no vested right in being able to have your existing way of doing business perpetuated." He had made similar objections before, and the judge had, by and large, overruled them. Today Ferguson seemed to be in a different mood. "Any detriment that the Betamax may have upon Gillette," he said sternly, "any detriment that Betamax would have upon the advertising industry, any detriment that Betamax may have upon the manufacturers of products who wish to advertise on television, is immaterial in this litigation."

When the man from Gillette tried to get in a few words edgewise, Ferguson interrupted to explain that judges had to avoid becoming "over-awed by your power. Lawsuits," he said, "have to be confined to very narrow, specific issues, and in this case Gillette is not one of those issues."

The next morning *Daily Variety* reported that the judge had "thrown a big wet blanket over the broader implications of the Betamax trial." But Kroft refused to put such a dark interpretation on Ferguson's rulings. "What the judge has said," he valiantly told *Variety*'s trial reporter, "is that we've given him our message loud and clear and he doesn't need extra witnesses."

The next ruling threw a wet blanket on *that* theory. When Dunlavey rose to offer the customary motion to dismiss with which the defense half of a trial is wont to commence, Judge Ferguson denied it but added that his denial should not be taken as a reflection on the merits of the motion. "The case is a big case and it is a case that will certainly go to the Court of Appeals," he said. "I am certain that a petition for certiorari will be presented to the Supreme Court. It just makes sense in the administration of the system that the Court of Appeals and the Supreme Court, when they are deciding on the issues before them, have the benefit of a full and complete record."

Now Kroft had to listen as Dunlavey presented his own survey statistics, his own documents, and a series of witnesses representing copyright interests with no objection to home taping—among them the baseball, football, and basketball leagues; the National Collegiate Athletic Association; the National Association of Religious Broadcasters; the Board of Education of the City of New York; and Fred ("Mister") Rogers. Dunlavey had also obtained, and introduced into evidence, an in-house study of the potential consequences of videotape recording commissioned by NBC in 1967. It concluded that the "effect upon commercial television and specifically upon our business would be minuscule." He quoted from a speech that Julian Goodman, the chairman of NBC, had made in 1978. There was no reason, Goodman had predicted, why video recording and broadcast television "cannot grow and prosper alongside of each other." And Dunlavey read from ten years' worth of documents purporting to prove an awareness at Universal, MCA, and Disney of Sony's work on home videotape recorders. "They knew," Dunlavey said. "They did not complain. They encouraged. Mr. Morita relied upon their encouragement."

Kroft responded by citing the sworn depositions of Morita and Schein, both of whom had acknowledged being warned against trying to sell a product like the Betamax. Far from being the innocent party depicted by Dunlavey, Sony "knew from as early as 1974 that there was a potential copyright problem," Kroft argued, and it had been "hiding its knowledge from the public."

This time the judge cut both lawyers short. He also nixed Kroft's request to call a rebuttal witness—an engineer, one Richard J. Stumpf, who was supposed to testify to the ease with which Sony, if it wanted, could make a videotape recorder that served all the legally innocent purposes of a Betamax and none of the culpable ones. Stumpf had conceived a system that could render a Betamax incapable of recording a program unless the broadcaster—presumably on the copyright holder's say-so—chose to let it be recorded. The system relied on a simple jamming device that could be installed in a Betamax at a cost, Stumpf was prepared to testify, of less than fifteen dollars a machine. Expert or no expert, Stumpf could not persuade Judge Ferguson that such a thing was workable—or relevant. If he were to order Sony to install a jamming device, "as sure as you or I are sitting in this courtroom today," Ferguson said, "some bright young entrepreneur, unconnected with Sony, is going to come up with a device to unjam the jam. And then we have a device to jam the unjamming of the jam, and we all end up like jelly."

In his closing statement Kroft took aim at Dunlavey's nonobjectors and his insistence that the Betamax served other purposes than the ones Universal was suing over. "That's just a lawyer's argument," he said. According to the the Field Research survey, the great majority of Betamax owners were taping movies and TV series, Kroft noted, not news shows or any of the other miscellaneous categories of programming that, according to Dunlavey, could be copied with impunity. "There is no evidence, nothing in the record," Kroft concluded, "to indicate that the defendants ever expected, before their lawyers went out and found these . . . witnesses, that a Betamax would be used to record off-the-air in a noninfringing way."

When Dunlavey's turn came, he reiterated his "lawyer's argument" and went so far as to claim that on its strength alone Sony was in the clear. "There never has been a case in history," he said, "where the manufacturer of a machine with a legitimate use was ever punished for some improper use made of the machine by a purchaser. . . ." From the waters of patent law he fished out something called the "staple article of commerce doctrine," which held that the supplier of a product used to commit patent infringement could not be implicated on that basis alone if the product had "substantial noninfringing uses." The Betamax, Dunlavey said, was a staple article of commerce. "What Mr. Kroft says is, 'Well, if you don't sell the Betamax, then we'll have none of these problems.' Well, I suppose you can't argue with that. If you kill the youngster, he won't grow up into a criminal." Perhaps the attorney for the plaintiffs had legitimate complaints about the selling and trading of tapes or the copying of cable programs in violation of contracts forbidding it, Dunlavey acknowledged. But if so, "Instead of using a shotgun, he should have used a rifle. He should have said, 'This particular man's use I object to, that particular man's use I object to. Would you please enjoin it?' But he has never tried to do that. He has gone for broke. He wants it all or nothing. He wants all recorders taken off the market." Universal's lawsuit, Dunlavey added, was "sort of like 'Tar Baby'—every time you hit it, you get stuck to it."

Judge Ferguson praised the lawyers for their "exceptional" preparation and for "making the trial pleasant for the Court." It had all been quite civilized. "I don't remember a single outburst—a single moment of acrimony," Jim Harwood, who covered the trial for *Variety*, said later. "Everybody wanted to show how professional and high-priced they were."

The decision came down in October. "Home-use recording from free television is not copyright infringement," Judge Ferguson wrote, "and even if it were, the corporate defendants are not liable and an injunction is not appropriate." Fair use applied, he decided, for two reasons: because the taping at issue was done by individuals or families in the privacy of the home and because Universal and Disney had "voluntarily" transmitted their works over the "public airwaves."

AS SACRED AS WHISKEY

A COPYRIGHT LAWYER displays many of the general characteristics of the legal species. He knows the difference between habeas corpus and corpus delecti. He can cite a precedent. He can write a threatening letter. He may well do business out of the same offices as specialists in tax law, labor law, and the other standard areas of legal expertise, and from time to time his duties may even require him to consult with these colleagues. But when he is behind closed doors working on a brief, or catching up on the latest court decisions, or explaining a complicated case to a friend, it is the rare copyright lawyer who does not regard his field as an airier and more splendid region than the rest—an intellectual Shangri-la in which he has had the great good fortune to settle and take out citizenship papers.

Few students start law school with thoughts of copyright. The typical copyright lawyer came upon the field by chance and developed a fondness for it. His susceptibility may owe something to youthful dreams of a career in show business—a common zone of fantasy life among future lawyers. In any case, he will have *chosen* to be what he is rather than have been led there by a boss's orders or a calculation of where the most money can be made. And he will earnestly believe that copyright is one of the necessities of civilized life.

Ask a copyright lawyer what sets his breed apart, and he may well answer as Stanley Rothenberg, a thirty-year veteran with the New York firm of Moses and Singer, does: "You have more fun. It might be reading a story and then looking at a film or a play to see whether it was derived from the story, or looking at a greeting card to see whether it comes too close to a poster, and then asking, Was it simply done in the same style or the same school, or was it a misappropriation of what was original with the first article? It's something you can share with your family. I was working on a case that involved a science-fiction story and a television series, and I had my two sons read the story and watch the program and decide for themselves if the one was taken from the other."

David Lange was a very junior member of the faculty of the Duke University Law School when, in 1969, he was asked to teach a course in copyright, patent, and trademark law. He enjoyed the subject so much that he wound up devoting his career to it. "It's one of the few areas of law, at least for me, that has ever gripped my attention and held it unremittingly," Lange explained. "Most property law applies to something that you can experience with your senses. The really neat thing about intellectual property is that it's entirely conceptual. It's like a constantly shifting game of three-dimensional chess in which you can't see the pieces."

The law of copyright had its origins in an understanding between the English printing trade and the crown. United by a desire for order and a distaste for the Scotsmen and other unreliables who periodically sought to elbow their way into the publishing business, they established a structure of censorship and licensing laws which endured through much of the sixteenth and seventeenth centuries, eventually giving a small society of printers—the Stationers' Company—control over "all the literature of England," to quote Lord Camden, an eighteenth-century jurist and legal commentator. It was a simple exchange which served both parties. The stationers surrendered editorial freedom (not only for themselves, of course, but for the entire British nation) in return for a grant of economic monopoly. But after Parliament let the arrangement lapse in 1695, piracy flourished, and the stationers, according to Lord Camden, appeared "with tears in their eyes, hopeless and forlorn" and "brought with them their wives and children to excite compassion, and induce Parliament to grant them a statutory security."

In addition to their wives and children—and for much the same reason—they brought authors into the picture. A publishers' monopoly was no longer a politically acceptable purpose, so the stationers based their plea on an author's right to the fruits of his labor. The upshot was the Statute of Anne, enacted in 1710, which gave English authors, for the first time, a power over their writings that went beyond ownership of a manuscript and the right to sell it outright to a publisher, for whatever the market would bear. Ever since, publishers (and record companies, movie studios, and other duplicators and distributors of copyrightable works) have pressed their interests by invoking the interests of authors, although like military contractors discussing national security, they have learned to accept a degree of suspicion of their motives as a cost of doing business. And ever since, the cause of copyright has had to contend with fears of monopoly and censorship, much as the cause of defense has had to contend with fears of militarism and war.

The Statute of Anne had all the basic elements of modern copyright: recognition of a work of art as intellectual property, a statement of the author's powers over it, and a fixed term of years before the work became fair game. By the end of the century the concept had acquired enough clout to earn a niche in the United States Constitution. Article I, Section 8, Paragraph 8 gives Congress the power "To promote the progress of science and useful arts, by securing for limited times to authors and inventors the exclusive right to their respective writings and discoveries." But the right was still very narrowly defined by today's standards. The first American copyright statute, enacted in 1790, covered only books, maps, and charts and only the printing, reprinting, or importing of virtually exact duplicates. Through a good part of the nineteenth century anyone was free to abridge, to translate, to dramatize, and even to correct a copyrighted work—all on the theory that, as Robert Grier, a Supreme Court justice from 1846 to 1870, observed, there could be "no hybrid between a thief and a thinker."

The law also refused to recognize any hybrid between an author and a foreigner. The idea of reciprocal copyright agreements among nations took hold in France in 1810 and in Britain and most of Germany in 1837. But Americans adopted the convenient view that the progress of *their* useful arts could best be promoted through unrestricted access to everybody else's. In the 1830s and 1840s, New York newspapers began publishing broadsheets or "supplements" with English and French novels reprinted in many columns of eye-straining type. So great was the demand for these works that some publishers sent swift boats out to meet the ships carrying the latest novels from England and did the reprinting on board during the return trip. When the *Evening Tattler* managed to obtain and reprint the first installments of *Nicholas Nickleby* a few weeks before its rivals, the editors trumpeted their achievement as hot news and self-righteously promised that their London agent would continue to supply, "by the big ships, all of the original sayings and original productions that are indigenous to that capital." But their commitment did not extend to a willingness to pay the author for his work. And while Charles Dickens was able to secure some justice, ultimately, from reputable book publishers, who paid for advance sheets and the right to print "authorized editions," his tireless campaign to change the law went unrewarded during his lifetime, and his views of the American press and Americans in general never quite recovered from those early encounters.

The situation was equally vexing to American writers since they had a terrible time trying to sell their works to American publishers who "had the best of Europe available for little or nothing," as a recent historian of copyright, Albert J. Clark, has written. The absence of an international copyright agreement not only cost writers money but, in Clark's view, stunted the development of American language, literature, culture, and self-respect. It put the country under the thumb of a form of English tyranny that persisted long after the revolution—a tyranny that provoked James Russell Lowell to write:

> You steal Englishmen's books and think Englishmen's thought;
> With their salt on your tail your wild eagle is caught;
> Your literature suits its each whisper and motion,
> To what will be thought of it over the ocean.

Mark Twain, with his enormous popularity in England and Europe, demonstrated that Americans could be significant producers as well as consumers of literature. Twain found English publishers who would pay him, but in the absence of copyright relations between the United States and Great Britain, he couldn't keep the nonpayers from the field. He couldn't even keep one of them, John Camden Hotten, from putting Twain's stories into the same books with other authors'—including Hotten's own—and giving Twain credit for the lot. "My books are bad enough just as they are written," Twain complained in a letter to the *Spectator* in 1872. He suggested that the title pages of Hotten's volumes ought to carry "a picture of a man with his hand in another man's pocket, and the legend 'All Rights Reserved.' " Twain, Bret Harte, Joel Chander Harris, John Greenleaf Whittier, Walt Whitman, and Louisa May Alcott put their names to an international copyright petition in 1886, and Twain was allowed onto the Senate floor to lobby personally for the cause. He told a Senate committee that he looked forward to the day "when, in the eyes of the law, literary property will be as sacred as whiskey, or any other of the necessities of life." But another five years went by before Congress was moved to pass a bill.

It was the clout of publishers rather than authors that carried the day. The Civil War gave rise to the first paperbacks—ten-cent editions similar to today's mass-market paperbacks in size and shape—and in the postwar years low postage rates and pirated English novels helped sustain the form. In the same period there emerged the Scribners, Houghtons, Lippincotts, and Putnams—patrician publishers who dealt in quality editions, believed in paying authors foreign and domestic, and thought that the competition should obey the same rules they did. Their view of the paperback threat was summed up by *Publisher's Weekly* in 1884: "In the rage for cheapness, we have sacrificed everything for slop, and a dainty bit of bookmaking is like a jewel in a pig's snout." An alliance of publishers, authors, and intellectuals fought relentlessly for an international copyright law and gathered momentum in defeat until Congress finally succumbed in 1891. Even then the morality of the thing was not very persuasive; before Congress would swallow the idea, it had to be sandwiched into a protectionist bill which, as a condition of copyright, required all books by American authors to be wholly manufactured in the United States—a rule that remained in force until 1986.

From the mid-nineteenth century on, advancing technology inspired new forms of expression, and one by one they claimed a place in the rubric of copyright. If a lithograph could be protected, why not a photograph? If a song, why not a player piano roll? If a play, why not a movie? If a dictionary, why not a computerized data base? As more works became eligible for copyright, the powers associated with it also expanded. If it was wrong to take someone else's creation as a whole and market it as one's own, wasn't it also wrong, people began to ask, to take a significant part of that creation—a plot, a character, a scene, a premise? As Congress and the courts took up these issues, they claimed merely to be applying established principles to new circumstances, and many of the steps they took, nudged along by technology and analogy, seemed short and straightforward. But cumulatively a great distance was covered.

In 1853 a federal court told Harriet Beecher Stowe she could not prevent—or demand compensation for—an unauthorized translation of *Uncle Tom's Cabin* into German (for German-American readers). In 1936 the United States Court of Appeals for the Second Circuit ruled that a Metro-Goldwyn-Mayer film *(Letty Lynton),* although based on actual events, violated the copyright of a play on the same subject by using a few of its interpolations of plot and character. In 1962, in another Second Circuit case, the court decided that a plastic Santa Claus manufactured by the Ideal Toy Corporation was a work worthy of copyright protection and that the Sayco Doll Corporation's Santa bore an infringing resemblance to it. Clearly something had happened to copyright law over the years. It had evolved from an extremely limited grant of powers to a limited class of people into a kind of homestead act for almost anybody engaged in an activity that could remotely be called artistic. Whoever carved out a piece of territory first—or first in the eyes of the law—got the right to develop it, alter it, and keep trespassers off. The law's change was, to some extent, a belated reflection of a cultural change—a growing reverence for originality and scorn for imitation. But the willingness to protect commercial art and photography (the latter widely regarded, in its early years, as a crude mechanical business unworthy of the name art) suggested that another, stronger influence was operating: an awareness of the huge sums of money involved in some of the new areas of technology. In 1912, when Congress gave copyright protection to motion pictures, it explained that it was doing so because "the money invested therein is so great and the property rights so valuable."

Throughout these changes the Anglo-American approach to copyright was thought to be fundamentally different from the approach taken by France and other European countries. There copyright was a natural right—akin to, say, freedom of speech. Here, as Congress declared in a preface to the 1909 Copyright Act, the law was "not based on any natural right that the author has in his writings . . . but upon the ground that the welfare of the public will be served and the progress of science and useful arts will be promoted." The amount of protection given, then, was up to the discretion of Congress and ought to depend on two factors: "First, how much will the legislation stimulate the producer and so benefit the public, and second, how much will the monopoly granted be detrimental to the public?"

But this may have been a case of "The Congress doth protest too much." The distinction between the natural right and the incentive views was never clear-cut, and it grew fuzzier with the years. As legal concepts will, copyright gained substantiality in the minds of lawyers, judges, and, above all, copyright holders. It became a thing unto itself and a thing that—such is the ingenuity of the legal profession—could be bought, sold, bequeathed, or carved into pieces, as the situation required. People began to speak of the "sanctity of copyright" as they spoke of the "sanctity of private property." Increasingly the law's judgments tended to turn on a determination of what was "fair to the copyright holder" rather than what would stimulate authors to express themselves, and the courts took care to see that one form of creativity received the same respect as another, even if a whole new body of law had to be developed for the purpose. In 1972 a California court (in a decision later overturned on a technicality) ruled that the right to market dolls and other products that played on a resemblance to Bela Lugosi as Count Dracula rested with the actor's widow and son rather

Century Park West Garage

Receipt Ticket

Entry Exit

04 05904 0 12MAR 2126 10 0408 13MAR 0008 125 003.

than with Universal Pictures, which had produced the movie *Dracula* in 1931. The judge didn't say his decision was intended to spur anybody's creativity. He simply pointed out that an actor, through the medium of his face, voice, and body, could create something that deserved protection as much as a book or a movie. In more recent applications this concept of a "right of publicity"—or a "human copyright," as it has been called—has enabled the estate of Lester Flatt (of the banjo team of Flatt and Scruggs) to halt the use of his likeness in ads for Coors beer and Johnny Carson to prevent the sale of a portable toilet called Here's Johnny.

Copyright lawyers, like all lawyers, enjoy a good argument, and as intellectual-property law reached into new realms, they often debated whether in this case or that the courts had gone too far. But the overall pattern of the law's expansion and elaboration, and the increasingly close scrutiny given to such issues, struck most people in the copyright community as a healthy thing for society—and it was certainly a healthy thing for copyright lawyers. In the 1960s and 1970s, however, their joie de vivre was dampened by a feeling that the formidable system of defense they had so painstakingly constructed was being overrun and that the world at large did not much care.

Printing, the technology that brought copyright into being, was an activity requiring large-scale facilities that gave the enforcers of the law a point of control. The old-fashioned copyright pirate was, therefore, reasonably easy to find and bring to justice. The modern pirate came equipped with new weapons: audiocassette recorders, compact photocopiers, personal computers, and other machines that transformed the act of copying from a cumbersome, expensive, conspicuous business into a simple, cheap, widely dispersed one.

The impatience of a physician named William Pasanoff, who had gone into the business of publishing medical journals, provoked the case that gave copyright people the most grief—at least until *Universal* v. *Sony* came along. Libraries had got into the habit of pooling resources when it came to periodical subscriptions. Library A would specialize in certain journals, knowing that Library B or C would supply it with whatever publications it lacked. From that habit they had graduated to dispensing *copies* of articles rather than the originals. "Interlibrary loan" the process was called, but the "loan" part was a euphemism since no one was expected to return anything. The publishers of high-priced technical journals like the ones that Pasanoff's company, Williams and Wilkins, dealt in were appalled by this system. Pasanoff's own distress focused on the fact that his primary readers—doctors—could get virtually free copies of articles in his journals through two federal government institutions, the National Institutes of Health and the affiliated National Library of Medicine. Nor was this an obscure service utilized by a select few; between them the NIH and the NLM were churning out hundreds of thousands of documents, running into millions of pages, every year. It seemed to Pasanoff that his own government was trying to run him out of business.

In the library world there was a certain amount of sympathy for these concerns, and negotiations were under way in the hope of establishing guidelines that would be acceptable to everybody involved. "The copyright community is famous for trying to work out agreements that avoid these touchy issues, and they really didn't want this case to be brought," Alan Latman, Pasanoff's lawyer (and a noted copyright authority who, at the time of his death in 1984, held the Walter Derenberg chair in intellectual property at New York University), said later. "But Bill Pasanoff is a very cantankerous M.D., and he decided he couldn't wait any longer and brought the case in 1968." When the trial court ruled his way, Latman's immediate thought was to push for an informal resolution. He and his client were willing to settle for a relatively modest scale of payments from the two government libraries in order to establish the principle of compensation. "Flushed with victory," Latman recalled, "we were on the plane the next day to Washington to make an offer of a blanket license that the government couldn't refuse—but did." The government not only refused but appealed and won a reversal from the Court of Claims, which, stressing the importance of medical research, decided that the practices Pasanoff objected to fell within the bounds of the doctrine of fair use. One dissenting judge characterized the opinion as the "Dred Scott decision of copyright law."

The case "died with a whimper," in Latman's words, when the Supreme Court deadlocked on it, four to four, with Justice Harry

Blackmun abstaining, apparently because of a conflict of interest stemming from his prior service as general counsel to the Mayo Clinic. This left the ruling with the status of law as far as Pasonoff and the Federal libraries were concerned, but without value as a precedent. In the 1976 Copyright Act Congress softened the blow by declaring that "systematic photocopying" by libraries was infringement. Unfortunately for the publishers involved, the act failed to explain what made copying systematic. "Back to the negotiating table" was how copyright people read the message Congress had sent them.

While the outcome of *Williams and Wilkins* made many copyright people unhappy, they could take consolation in the view that it was a narrow opinion—anchored to the field of medicine—and that it had been overridden by Congress. The district court decision in *Universal* v. *Sony* was more disheartening. For the first time a court had ruled that copying for mere entertainment or convenience was fair use, that copying of a whole work could qualify, and that individual copiers could gain immunity from the law simply by committing their offenses in a noncommercial setting, while frankly commercial interests—the suppliers of the machines involved—could profit as much as they liked. Finally, and most ominously, the decision suggested that infringement wasn't really infringement unless the victim could prove economic harm.

Harm was an extremely elusive and speculative issue. Since no one could predict how much money a particular book, movie, or record would make, how could anyone show to what extent—or even whether—an act of infringement had cut into that potential? Many copyright lawyers believed that the law really shouldn't consider harm except as a guide in assessing damages. Making harm a factor in determining whether infringment had occurred, it seemed to them, implied that a copyright owner had to be in financial trouble before he could exercise his legal rights. The federal register of copyrights, David Ladd, gave a lecture on this theme entitled "The Harm of the Concept of Harm in Copyright." " 'Harm' as a basis for policy," said Ladd, "substitutes, to borrow a phrase from welfare legislation, a means test for the historic and honorable principle of just compensation measured solely by what the public chooses to pay. . . ." Although Ladd didn't cite the Betamax case by name, he seemed to have it in mind as he defended the need to respect the rights of copyright holders even when they appear to be "well-off enough that enhanced protection will only mean windfall profits. . . . The glory of copyright," said Ladd, "is that it sustains not only independent, idiosyncratic, and iconcoclastic authors, but also fosters daring, innovative, and risk-taking publishers. . . . Copyright supports a system, a milieu, a cultural marketplace which is important in and of itself. . . . It does not 'give' the author or the publisher anything. It cloaks in legal raiment the undoubted right. It does not guarantee success, or audience, or power, or riches. It is not a warranty, but an invitation to risk. When the rewards are large, we should not resent or envy, but rejoice, and we should likewise cherish every miserable failure."

Universal v. *Sony* had drawn a certain amount of public attention, involving, as it did, powerful and glamorous industries and bearing, as it threatened to, on the everyday behavior of ordinary Americans. After the district court handed down its decision, however, the press generally regarded the issue as settled, and so did most people with a professional interest in it. By the middle of 1981 many casual followers of the case probably assumed that the decision had long since been upheld since that would have been a small story for the back pages, and they could have missed it. Only the immediate parties to the lawsuit, a few of their business allies, and a few hundred copyright aficionados were still waiting for a follow-up, and only they were at all prepared for the decision handed down by the United States Court of Appeals for the Ninth Circuit on October 19, 1981—or "Black Monday," as it was promptly designated on Sony's calendar.

The appeal had been argued a year earlier, in San Francisco, before three judges: John F. Kilkenny, a Nixon appointee who was a conservative in style as well as politics; William G. East, a senior judge with the trial court in Oregon who was a fill-in at the appellate level; and William C. Canby, Jr., a Peace Corps veteran, the brother-in-law of former Vice-President Walter Mondale, and, at fifty, the junior member of the panel by a margin of more than twenty years. *Universal* v. *Sony* was the last case on their schedule that day, after one involving a Nevada casino. Dean Dunlavey felt that the judges had paid extra-close attention to the casino case, owing to the attractiveness of one of the attorneys, and had rushed through the Betamax argument in order to make up for lost time. "It was dry and boring," Dunlavey said later. But one of the judges asked him a question that Steve Kroft found encouraging. "He pointed out that motion pictures were given extra-special treatment in the copyright act, and he asked Mr. Dunlavey what was his response to that," Kraft said. "I thought that was a pretty incisive question. It was an indication that he was taking our position seriously."

Dunlavey came to the same conclusion after a few months had passed and the Ninth Circuit remained silent. To affirm the findings of the trial court ought not to be a particularly time-consuming task, he reasoned. "It was obvious it was going to be something out of the ordinary, as indeed it was."

The decision sided with the plaintiffs across the board. Speaking for a unanimous court, Judge Kilkenny found that home video recording was infringement and that Sony was responsible for it. As *Video Week* reported, "Ruling was complete surprise to all but readers of legal journals that have carried a number of articles criticizing Ferguson decision."

Judge Kilkenny chastised Judge Ferguson (who was now his equal, having been elevated to the appeals court after the Betamax trial) for failing to understand that the copyright act was a broad statement of rights limited by strict and specific exceptions and for stretching fair use beyond the criteria set forth by the statute—criteria that the appeals court regarded as clearly inapplicable to videotaping in the home. If the statute had nothing to say about home copying, its silence was to be taken not as a sign of ambiguity but as a reason to apply the general terms of the act without qualification. "Our study of the record and analysis of the legislation," the opinion continued, "convinces us that the district court inadvertently bypassed the statutory framework of the 1976 legislation. . . ." The Ninth Circuit conceded some uncertainty about the audio-recording question but noted, "The copyright statute treats sound recording and audiovisual works as separate categories of protected material. And Congress has shown special solicitude for audiovisual works."

On the harm question, the Ninth Circuit held that the lower court had been "much too strict" in requiring the plaintiffs to demonstrate the likelihood of harm. They had only to show that home video recording was likely to affect the market for Universal's and Disney's products, and that, according to the Ninth Circuit, "seems clear." Home video recording failed to satisfy the fair use test on other grounds: It involved entire works rather than portions of them, and its purpose was far removed from those served by the usual fair use activities. "It is noteworthy," Judge Kilkenny wrote, "that the statute does not list 'convenience' or 'entertainment' or 'increased access' as purposes within the general scope of fair use."

Dunlavey's nonobjectors and his "staple article of commerce" defense did not impress the Ninth Circuit at all. "Videotape recorders are manufactured, advertised, and sold for the primary purpose of reproducing television programming," the decision observed. "Virtually all television programming is copyrighted material. . . . That some copyright owners choose, for one reason or another, not to enforce their rights does not preclude those who legitimately choose to do so from protecting theirs."

The court speculated that Judge Ferguson might have ruled as he had partly because of the "difficulty of fashioning relief." The Ninth Circuit did not order an immediate halt in the sale of Betamaxes. It sent the question of relief back to the district court (although any further action would be stayed until the Supreme Court decided whether or not to hear the case). Looking forward to these deliberations, however, the Ninth Circuit cautioned the lower court not to be "overly concerned with the prospective harm to appellees" because "a defendant has no right to expect a return on investment from activities which violate the copyright laws."

The Ninth Circuit's decision vindicated the judgment of copyright people from coast to coast, and they took to the pages of the nation's law journals to celebrate it as a return to "basics" and a reaffirmation of the "traditional" idea of fair use. "Every other article we received over the transom was on the Betamax case," one of the people responsible for the *Journal of the Copyright Society* of the United States recalled. At the society's regular luncheon and dinner seminars in New York, she added, "All the speakers wanted to tie the Betamax case into their remarks whether it had any relevance or not. After a while people groaned when somebody got up and talked about the Betamax case."

"There was a feeling," Stanley Rothenberg said later, "that this decision would serve as protection against still further technological advances that would be even more damaging—whether it was going to be the personal Xerox machine that everyone would put in his home or the use, by organizations, of transmission devices with instant printers so that a large corporation could purchase one copy of a magazine and transmit the important articles to its divisions and its subsidiaries across the country instantaneously. And certainly there were a lot of developers of computer software who were concerned that the new devices that were being marketed would result in their programs being duplicated willy-nilly. It also meant that the recording companies might have a claim against the manufacturers of the equipment that enables people to put in a prerecorded cassette and a blank cassette and in a minute make their own duplicate. So the Ninth Circuit decision was very encouraging to the record industry. It was encouraging to the copyright community generally in regard to the contributory infringer—the one who makes the device whose primary function is to duplicate copyrighted material."

At firms like Rothenberg's, copyright people faced some skepticism from their noncopyright-absorbed colleagues. "Partners of mine would ask, 'How are you going to police it?' " Rothenberg said. "And the response was that first you have to establish the principle that the copyright owner's rights are being invaded and then the legislature can fashion an appropriate means of compensation."

DIVORCE, JAPANESE STYLE

FIVE YEARS HAD passed between the filing of the lawsuit and the Ninth Circuit's decision. The conversation piece in William Griffiths's living room had acquired an acronym—VCR—and some three million Americans had bought one. Not since color TV had a gizmo that cost so much sold so well. Not even MCA had been able to keep the dreaded device out of its corporate culture. A murder suspect in an episode of "Columbo" claimed, as an alibi, that he had been home watching TV, and to prove it he described the program he had been watching. Lieutenant Columbo brought him to bay by proving he had seen the show on a VCR *after* he had done the deed.

Sony could take a large share of the credit for the VCR's success. The company's reputation for excellence and reliability had helped consumers overcome their fear of the unknown (especially the expensive unknown), and in the United States an ingenious ad campaign had made the name Betamax almost as commonplace as Kleenex and Xerox, two other brands ending in *x*. Sony should have been sitting on top of the world. Instead, it had been thrown on the defensive in a global popularity contest with a competing version of the same product—one that had appeared a year and a half after the Betamax—and the Sony share of the VCR market was falling at an alarming rate.

If it were necessary to give a one-word explanation of the change in Sony's fortunes, the word would be *Matsushita.* Few Americans have heard of the Matsushita Electric Industrial Company, although most have probably pushed a button on one of its products, which include virtually every communications device and home appliance known to man, and a good deal else besides. In Japan, Matsushita Electric ranks second to Toyota Motors in sales—at last sighting, it stood twenty-fourth on *Fortune* magazine's list of the world's largest industrial corporations—and its founder, Konosuke Matsushita (the accent is on the second syllable of both names) regularly places high on polls of "People I most admire."

In 1917 he was a twenty-two-year-old inspector with the Osaka Electric Light Company and a young man so filled with enthusiasm for the promise of electricity that when his employer showed no interest in a device he had fashioned in his spare time—a piggyback adapter that allowed an electric appliance to be plugged into a light bulb socket, the sole source of electricity in most of the electrified homes of that era—he quit and began to manufacture the thing on his own, with a pool of capital drawn from his savings, a loan from a friend, and the pawning of his wife's kimonos. The adapter was an archetypal product in Matsushita Electric's history just as the tape recorder was in Sony's. Not remarkably ingenious or novel, it would "never be mentioned in the same breath as Thomas Edison's lightbulb," an article on the company in the *Wall Street Journal* observed. But Konosuke Matsushita made his version more convenient to use than others (by adding a rotating collar on the plug end of the extension cord that came as an accessory), and he undercut the competition's prices by thirty percent (by building the screw-in element from discarded light bulbs).

Efficiency in the use of labor as well as materials became a basic element of the Matsushita creed. Konosuke Matsushita belonged to a generation of Japanese industrialists who were deeply influenced by the efficiency theories of the American Frederick W. Taylor, whose 1911 book *The Principles of Scientific Management* sold more than two million copies in Japan. Taylor's preoccupation with keeping workers busy grated on the sensibilities of liberal readers in Britain and the United States, but his ideas were well suited to the paternalistic setting of Japanese industry and well timed to influence Japan's relatively belated—and abrupt—industrialization.

In 1923 Matsushita came out with a battery-powered bicycle lamp that ran for thirty to fifty hours—ten times the life of any bicycle lamp then on the market. Sales went slowly at first because the public had decided that gas- and candle-powered models were more reliable. "Our potential customers were simply not prepared for our long-life lamp," Matsushita wrote later. ". . . They gave us no credibility at all." His solution was to give away ten thousand free samples to retailers, who were asked to turn them on and leave them in their windows. The Matsushita bicycle lamp became the people's choice.

Matsushita Electric went into the radio business in 1930. Other Japanese companies were already making radios, but they were delicate and fault-ridden instruments which had to be sold through specialized dealers capable of explaining and, if need be, repairing them. Konosuke Matsushita decided that *his* radios would be simple and fault-free so that anybody could sell them, and the prevention of defects became a company obsession.

Until the thirties it seemed to him, as it seemed to most Japanese entrepreneurs, that the pursuit of profits was reason enough for a company's being. In 1932 the sight of a vagrant filling a tin cup from a public fountain caused him to rethink that idea. "Although the water had to be processed and distributed, it was so cheap that nobody had to think twice about using it," he explained later. "I began to think about abundance. And I decided that the task of the industrialist is to make his products widely available at the lowest possible cost to bring better living to the people of the world."

He propounded this philosophy in a speech to his assembled employees—all 175 of them. Half a century later Matsushita Electric had 150,000 employees spread among eighty-three manufacturing and sales subsidiaries in thirty-seven countries, and the thoughts of the chairman (or the executive adviser, his title from 1973 on) continued to hold great sway. Always a frail man, Matsushita had battled tuberculosis in his youth, and he had seen several brothers and sisters die of it. In later years he could often be found in a permanently assigned room at the company hospital. But he turned ninety in November 1984, a year when he had the additional distinction of being Japan's largest individual taxpayer, contributing over nine hundred million yen (more than three and a half million dollars) to the public coffers. A living advertisement for the benefits of chronic ill health, he acknowledged the honor with a statement from his hospital bed. "Taxes are too high," he said.

By then his public appearances had become rare, and his once-powerful voice had grown thin. A slight, birdlike figure, he was not easy to make out behind a lectern, but however softly or briefly he spoke, the audience would listen raptly. And for employees who failed to hear him in person, there was the option of joining a study group to discuss his essays or listen to cassettes of his talks on such subjects as the importance of quality control, the need to learn from adversity, and the "new age of prosperity" in the offing for Japan. Konosuke Matsushita took up this theme in the late 1970s, when a lot of people—foreigners in particular—had the impression that Japan had already entered an age of prosperity. For thousands of years, he observed, economic vitality had been shifting from region to region in a westward direction, from Egypt and Greece to Rome, Western Europe, and North America, "following a natural flow of history." And it followed that "in the twenty-first century . . . the center of prosperity will be in East Asia. . . ."

Before his retirement Konosuke Matsushita practiced a light-handed style of management, and it filtered down through the ranks. The *sunao* mind—a heightened state of intellectual humility and clearsightedness—is the most highly cultivated quality at Matsushita Electric. The typical workday begins with a morning meeting that may include the singing of the company song ("Love, Light, and a Dream") and the recitation of the company creed ("Through our industrial activities, we strive to foster progress, to promote the general welfare of society, and to devote ourselves to furthering the development of world culture") or, alternatively, the seven objectives ("Spirit of service through industry; spirit of fairness; spirit of harmony and cooperation; spirit of struggle for the sake of progress; spirit of courtesy and humility; spirit of assimilation; spirit of gratitude"). In contrast to Sony, which has become a haven for the outspoken, Matsushita has produced a leadership corps noted for an air of modesty that, to an American sensibility, borders on obsequiousness. "And the higher you climb in the hierarchy," according to one middle-level executive, "the more modest you become." The company has even taken a modest attitude toward the promotion of its name, selling products under an assortment of brands—chiefly National (in Japan), Panasonic, Technics, and Quasar. Unlike Sony, it has been only too willing to make what are called OEM deals (the initials stand for original equipment manufacturer), in which one company puts its brand on another company's product. Matsushita operates on the theory that a product sold is a product sold, whoever's label it carries. As Richard Tanner Pascale and Anthony G. Athos have written in *The Art of Japanese Management,* "Matsushita, inspired by Henry Ford's pricing strategy for the Model T, grasped the concept of aggressively pursuing market share, gaining economies through manufacturing experience, and lowering price, thus establishing barriers to entry for competitors who found the small margins unattractive."

In 1933, when poor health made it hard for Konosuke Matsushita to keep to his regular schedule of meetings with his subordinates, the company adopted a divisional structure. Each division (originally there was one for radios, one for batteries and lights, and one for wiring, synthetic resins, and heating appliances) "would assume responsibility for planning all sales programs, production of new commodities, and even accounting. Indeed, each division would operate like a separate company."

When a Matsushita division needs money, it applies to the parent company for a loan, and if the application is approved, it pays interest just as if it had borrowed from a bank. If a division is doing well, it gets a large say in the disposition of its profits, and it needn't fear that its earnings will be raided to shore up the finances of another, ailing division. The divisions also enjoy striking autonomy in what might be called foreign relations—that is, dealings outside the Matsushita family. In the buying and selling of components, they have a mandate to cut the best deal they can, whether it means doing business with another unit of Matsushita or with a troublesome competitor. Sometimes Matsushita winds up competing with itself, as it did in the 1970s, when both the tape recorder and radio divisions marketed audiocassette recorders with built-in radios (or, if you will, radios with built-in audiocassette recorders). Divisional structures have become common in large, diversified corporations, but few have been as devoted to the concept—in adversity as well as prosperity—as Matsushita. Fully half the components it produces go into non-Matsushita products, which generally compete against Matsushita's own products. At one time or another it has supplied semiconductors, camera tubes, magnetic heads, tuners, displays, motors, batteries, and disc drives to, among other companies, Mitsubishi, Fujitsu, Hitachi, Toshiba, Sharp, NEC, Canon, and Sony. Matsushita components are known for precision manufacture as well as cost-competitiveness. Although the definition of a *robot* is shifty, Matsushita is widely regarded as the world's biggest maker of robots, and no assembly lines are more thoroughly robotized than its own. Its factories have a twenty-first-century look to them—a purring efficiency that makes as powerful impression on a visitor as do the morning meetings and the company song.

The Matsushita approach to product design has been called creative followership. "They don't like to be the pioneer," James Abegglen, a Tokyo-based management consultant, explained. "Pioneers have a lot of arrows up their asses. Matsushita likes to come in second, after the trial has been blazed a little and you can see which way is safest."

With a few exceptions, such as the Technics line of turntables, Matsushita products are rarely the last word in the field according to the electronics buffs. But they have a way of arriving just when the public at large is ready to consider buying a particular item and when the buffs have turned their attention to something else, and they frequently boast a lower price than competing products, a few extra features, or both. The buffs may not be dazzled. The competition usually is. As an American executive of Matsushita's Quasar subsidiary has said, "Their concept of research and development is to analyze competing products and figure out how to do it better."

Whatever its intrinsic strengths or weaknesses, a Matsushita product enters the Japanese retail world with a privileged status. Merely by virtue of its parentage, it is assured of shelf space in a franchise network of some twenty-seven thousand "National Shops" which sell Matsushita products either exclusively or predominantly. They constitute about forty percent of the electronics retailers in Japan. "That was a brilliant move by the old man in the fifties." Abegglen said. "He franchised appliance stores all over the country—mom-and-pop operations. It was an absolutely brilliant marketing scheme for Japan. You've got a little neighborhood shop. Matsushita comes along and says, 'Tell you what I'll do for you. I'll provide your sign, your advertising, your inventory, and you'll sell my products, won't you? If you sell eighty percent Matsushita products and twenty percent everybody else's, we'll give you a rebate. If it's ninety percent and ten percent, we'll give you a bigger rebate.' And that's the only contract he needs. What it means is that when Matsushita puts a new product on the market, it has maybe a twenty percent market share automatically."

It was partly in response to Matsushita's strong position in the Japanese retail market that Sony, in the late fifties and sixties, shifted its focus to the United States. There, too, Matsushita was hot on Sony's trail. By 1959 it had established an American sales subsidiary, the Matsushita Electric Corporation of America, known as MECA. By the end of 1960 it was making radios for Macy's on an OEM basis ("America's largest-by-far retailer joins hands with Japan's largest-by-far manufacturer of appliances," an ad in *Home Furnishings Daily* proclaimed), and it had two hot sellers: an AM/FM model that cost $79.95 and came with an auxiliary "concert speaker," a feature Sony had not yet thought to provide; and a six-transistor AM model that cost $29.95, or $10 less than the Sony equivalent. The zest for exports landed Matsushita in trouble in the early seventies. Shufuren, a federation of Japanese housewives, staged a boycott of Matsushita products, charging that some were being sold for less money in the United States than in Japan. After a sharp decline in profits Konosuke Matsushita announced a more evenhanded pricing policy and thanked Japan's consumers for calling the problem to his attention.

How do you compete with a company like Matsushita? The success of the U-matic suggested a possible rule of thumb for Sony: Approach carefully and with the hand of friendship outstretched. With the U-matic Sony had given the fruits of its R&D work away to both Matsushita and JVC, in return for a commitment by those companies to help establish an industrywide standard. From Sony's standpoint the deal could not have worked out more splendidly. Thanks partly to its good name, partly to the high quality of its machines, and partly to the fact that the U-matic technology worked particularly well with its Trinitron TV sets (and vice versa), Sony had captured an exceptionally high share of the market. With this happy experience fresh in their memories, Morita and Ibuka proposed in the fall of 1974 that the same alliance reunite behind the new Sony format known as Beta.

But at Matsushita and JVC there were people who did not look back on the U-matic experience with the same fondness—or look forward to the next adventure in technology sharing with the same eagerness—as the management of Sony did. The mid-1970s were a glorious period in Sony's history. Younger and smaller than most of its domestic competitors, the company enjoyed a worldwide reputation as an exemplar of Japanese high tech and a symbol of the country's leap from ruin to industrial eminence. Like many a child prodigy, however, Sony had developed unevenly. Its social skills were not as highly refined as its technical skills, and in what might be called affairs of the corporate heart, it retained some of the blundering quality of a child. Like many prodigies, too, it had charmed the press and the public in distant lands while leaving a trail of envy and enmity back home. The other parties to the U-matic alliance could not help noticing that while *they* regarded that format as a joint development by all three companies, almost everyone else regarded it as purely a Sony invention. And they could not help noticing that Sony had made a lot more money from the U-matic system than they had. If that was what happened when a company cooperated with Sony, maybe cooperation wasn't such a hot idea.

These concerns were felt with extra intensity at JVC, a company that, while financially tied to Matsushita, wanted no part of "followership," creative or otherwise. JVC—the Victor Company of Japan—was established in 1927 as a subsidiary of the Victor Talking Machine Company of the United States (which later merged with the Radio Corporation of America to become RCA Victor, which later dropped the "Victor" and became plain RCA). JVC began making phonographs and phonograph records in 1929 at a large plant in the port city of Yokohama, where the company still has its headquarters. In the late thirties, as Japanese-American relations deteriorated, ownership passed from foreign to domestic hands, although JVC never wavered in its loyalty to the Victor trademark, the dog Nipper with ears perked to "His Master's Voice." In 1945 American bombs nearly destroyed JVC's factory. Eight years later JVC was at bankruptcy's door and would have gone right on through except for Konosuke Matsushita, who bought a majority of the stock and placed the company under the Matsushita umbrella. Characteristically, though, he agreed to let JVC maintain a high degree of autonomy, and that became a point of almost obsessive pride with the company. Matsushita people would routinely refer to JVC as a subsidiary, but anyone who used that term with a JVC person was asking for trouble. "The only relationship we maintain with Matsushita is capital involvement—that's it," JVC executives would say.

Like Sony, JVC saw itself as an industry leader. It had produced Japan's first TV set (in 1939), LP record (in 1953), and stereo phonograph (in 1958). In the VTR field it had not enjoyed much commercial success, but it had come up with the world's first helical-scanning color VTR, a 1961 product, and it had contributed important ideas about color recording to the U-matic standard. And the ultimate goal of JVC's VTR work, as of Sony's, had always been a product for the home.

Almost from the moment the U-matic format was adopted, JVC began preparing to assume a more assertive role in the next generation of VCRs. In June 1971 Shizuo Takano, director of JVC's video products division, set out to develop a true home VCR. "There were only three of us, and we had to report to the top what we were doing," Yuma Shiraishi, one of Takano's colleagues, recalled later. The team had little to show for its efforts over the next two years, and Takano had to fend off suggestions that he drop the project in view of his division's sorry financial record. The main achievement of those years was drafting a "development matrix"—a set of specific goals. In the first JVC prototype the cassette had a recording capacity of one hour, and Shiraishi, looking over the TV schedule, noted that a dozen programs a day lasted more than an hour, so two hours became the recording time in JVC's matrix. He and his colleagues also benefited from lessons learned in the effort to sell U-matic-style machines as home VCRs—a campaign in which JVC persisted longer than Sony. The company built up such a large inventory of unsold machines—mostly the so-called "ensemble units," which combined TV and VCR and cost nearly 600,000 yen (or $1,670)—that twenty engineers were reassigned to sales. "They went out to the dealers, and they went out on door-to-door sales," Shiraishi recalled. "Every night, when they came back to the office, they would say, 'Well, today I heard this,' or 'Today I heard that.' " One thing they heard was that JVC's ensemble unit was much too big. The typical Japanese home is small and efficiently furnished. "Where would we place such a bulky machine?" people kept asking.

The energy crisis of 1973 had a brutal effect on the Japanese electronics industry, increasing production costs, decreasing demand, and creating a national movement against the production of luxury goods. But inflation eased the financial pressure on JVC's video division by making the existing inventory of U-format machines competitive again, and in a short period the company unloaded them all, mostly to the United States. In November of that year JVC named a new president, Kokichi Matsuno, and he and executive Vice-President Hirobumi Tokumitsu issued a challenge to the video division "to overtake and pass Sony."

Tokumitsu had been a superior officer of Morita's in the navy's technical division at the end of the war. When Sony demonstrated the Betamax for Matsushita and JVC in December 1974, Morita turned to Tokumitsu as an old friend and said, "JVC is well established in the VTR field. Let's continue to work together on home video as we have with professional video." But Tokumitsu was thinking: "Ours isn't finished yet, but it's better. In another six months we can really make it into something."

The picture quality on JVC's prototype was still seriously unsatisfactory, but the recording time had been increased to two hours, and among the leaders of the company and its video division there took hold a belief that if they played their cards right, JVC's system rather than Sony's might become the standard. The Betamax was nothing special, to hear JVC's engineers tell it. "If Betamax had been a revolutionary product," one of them said later, "we were ready to abandon our own development in the interests of a common standard. But to us it was only one of many tries by many companies. It was sort of like a miniature version of U-matic—the motor was actually the same—so we didn't think it was difficult to build. We didn't feel pressed to be in a hurry."

If the Betamax itself did not make a profound impression on JVC, Sony's attitude toward it did. "They didn't ask if we had any questions or if we had any opinions," one JVC engineer recalled. "They just asked if we would like to go with this. They were very confident. The overall impression that I got after attending the meeting was that we had no choice." The U-matic had been a "genuine joint development," he added. "Betamax, on the other hand, was really a finished product, with all the design and the size and everything, so there was no room for compromise. And then we heard from some component manufacturers that Sony had already tooled up. They had already prepared the molds to make the equipment." The JVC people read this to mean that Sony intended to go into production before anyone else and take a high market share—just as it had done with the U-matic. JVC's management concluded that "the 'sharing' of technology would obviously work in only one direction."

At Matsushita some people formed a similar impression. After the demonstration Matsushita's technical people met with their Sony counterparts and reported that they liked everything about the machine except the one-hour recording capacity. One of the Sony representatives replied, "It may be a problem, but we've already started tooling up, so it cannot be changed."

"Our executives knew that Sony had already completed its molds—its dies—and Sony wanted to sell the product the following year, 1975," Tsuzo Murase, the general manager in charge of video engineering at Matsushita, said later. "So it was almost like an ultimatum. There was no room for negotiation, no room for exchange of ideas. They were saying, 'This is it. We've completed this one. If you're interested, follow us.' With three-quarter-inch U-matic, the three companies—JVC, Matsushita, and Sony—got together and came to a single standard in a friendly way. This could not be the case with Betamax. Of course, we proposed a modification to expand Beta to two hours, but Sony didn't listen because their marketing schedule was already set."

Matsushita had reason to be concerned about the recording time question. Only a year earlier, after all, it had put a cartridge-model

VTR with a thirty-minute capacity on the market. AutoVision, as this machine was called, had been a humiliating failure—the kind that led to resignations and reassignments and mass discombobulation since an entire department with some twelve hundred employees had been involved. When the responsible executives tried to reconstruct where they had gone wrong, the thirty-minute factor loomed large. Like their colleagues at JVC, the top people at Matsushita's video division believed that a recording capacity of more than an hour was called for.

In Akio Morita's mind, however, an hour seemed sufficient. Questioned on the subject by an American subordinate, he replied that the one-hour capacity of the Betamax had been dictated by a more important consideration: the need to keep the cassette small. Eventually it might be possible to go beyond an hour, Morita said, but Sony could not afford to wait, lest some other company—Matsushita, for example—bring a home VCR to market first. When the Betamax went on sale in may 1975, a reporter asked Morita about an RCA market survey indicating the desirability of two-hour recording. Morita pointed out that the great majority of TV shows lasted an hour or less. Anyone who wanted to record a program that ran longer than that, he said, could simply use two cassettes.

Disregard for market surveys was an old Sony tradition, of course. Time and again Morita and Ibuka had made decisions that went against the wisdom of the industry, and time and again their confidence had been vindicated. By the mid-seventies, however, some people felt that Sony's attitude had passed into a zone *beyond* confidence. "The Betamax was a product that had a lot of Mr. Morita's ego in it," a Sonam executive said later. "We used to go to Japan twice a year, and I remember a product lineup meeting—it was over the Fourth of July weekend before the introduction of the Betamax deck—where we saw a unit with a built-in electronic clock in one of the product rooms, and Mr. Morita pointed to one of his engineers and said to me, 'I'm going to teach you something important in marketing. Look at this engineer. Do you trust him to make a videotape recorder? Yes. Do you trust him to make a clock? No, because he is a video engineer, not a clock engineer. So the clock should not be built in, because then if it breaks down, the entire unit has to go out for repair. We should have a separate clock that sits on top of the unit and looks as if it's part of unit.' And indeed, they changed the product and took the electronic clock out and put it on top, so that it plugged into the unit. I don't know if any clocks ever broke or not. I do know most of our competitors never saw the need for a separate clock, and they did all right."

The decision to put the Betamax on the market did not mean that Morita had lost hope for standardization. Sooner or later, he thought Matsushita and JVC would fall in line. Indeed, he could hardly imagine their doing anything else once they saw real-world proof of the Betamax's appeal. Morita's faith was shaken slightly, though, in June 1975, when Matsushita Kotobuki, a Matsushita subsidiary based on the southern Japanese island of Shikoku, began test-marketing an incompatible VCR, the VX-100. Matsushita executives assured Sony that their video products division in Osaka was still considering a number of home VCR options, including Betamax. The introduction of the VX-100, they said, was an independent move by Matsushita Kotobuki and did not necessarily reflect the parent company's inclinations. The VX-100 was certainly not very worrisome from a technological standpoint. It had been designed around a loading mechanism in which the drum entered the cassette (as opposed to the Betamax and U-matic approach, in which the tape is withdrawn from the cassette and pulled over to the drum). Simpler loading was the advertised advantage, but the concept resulted in a bulky cassette and made for problems in maintaining correct tape speed.

The protestations of the Matsushita management were true enough, as far as they went. The company was not committed to the VX-100, and the Betamax still had its champions in the ranks of Matsushita executives. Unbeknownst to Sony, however, the VX-100 was not the only alternative to Betamax that now figured in Matsushita's deliberations. By the summer of 1975 JVC's project, although still a secret from many of that company's own officials, had privately been demonstrated for a few key outsiders, including the top people at the video division of Matsushita. The Matsushita group found the JVC prototype interesting enough to propose a few modifications. At the time JVC was working with a half-flanged cassette, in which the two reels of tape, in effect, overlapped. Such an arrangement (which can be found in the standard audiocassette) is possible because both reels cannot be full at the same time, and it would allow JVC to record two hours on a cassette comparable in size to Sony's. But the half-flanged idea also created unnecessary friction since the bottom edge of the tape could not easily be protected from rubbing against the inside surface of the cassette. The Matsushita engineers recommended a full-flanged cassette to overcome this problem, and JVC eventually decided to follow their advice.

The JVC group also took Konosuke Matsushita into their confidence. Although semiretired, he was a powerful influence who might, at a critical moment, nudge his company in one direction or another. "It's marvelous—you've made something very nice," was his verdict.

The decision to approach him was partly a preemptive strike against Sony. Over the summer Konosuke Matsushita and Akio Morita held a series of wide-ranging conversations under the auspices of the PHP Institute, a kind of think tank (the initials stand for "Peace and Happiness through Prosperity") founded by Matsushita in 1946, when he had a lot of time on his hands as a result of the occupation forces' decision to remove him from the management of his company. (Matsushita Electric was to be dismantled as one of the *zaibatsu*—the monopolies that had dominated the Japanese economy before and during the war. Later, after much protest against the designation, the Americans relented.) Matsushita and Morita seemed to represent different viewpoints as well as epochs. One was from Osaka, a center of old wealth and traditional Japanese values. The other lived in Tokyo, a veritable melting pot by Japan's xenophobic standards, and a city built with comparatively new money. One, loosely speaking, was conservative, antigovernment, nationalistic; the other, liberal, politically aware, internationalistic. But in their discussions that summer, which formed the basis for a popular book about the future of Japan and Japanese industry, they found a number of things to agree on. To outsiders it seemed that something like a father-son relationship had developed, and it seemed reasonable to anticipate a period of close cooperation between their companies—a prospect that JVC, it is safe to say, did not relish.

In January 1976 Morita assumed the post of chairman of Sony, and

Ibuka, at the age of sixty-seven, become honorary chairman. "This year will be the first of the video age," Morita announced. He might equally have said, "The Betamax will be the first product of the Morita age." He had great confidence in it technically, and he was confident of its appeal to consumers once they had become accustomed to it. But the prospects for standardization seemed to be getting cloudier. Toshiba and Sanyo were still promoting their skip-field system, the V-Cord; Matsushita Kotobuki had come up with a new version of the VX-100, the VX-2000; and there were rumors about that JVC might have yet another incompatible format in the works. Sooner or later, Morita knew, Matsushita Electric would enter the home VCR market in a serious way. If it went with Betamax, the rest of the industry would almost surely follow suit. If it decided on some other system, there was no telling what would happen.

One of Morita's first act as chairman was to propose a summit meeting with Masaharu Matsushita, Konosuke's son-in-law and the reigning president of Matsushita Electric. (Masaharu Matsushita had sensibly opted to abandon his own name in favor of his wife's when they married.) They met on February 3, along with their seconds- and thirds-in-command, at Matsushita headquarters in Osaka. It was partly a get-acquainted session, but Morita used it to renew his plea for the Betamax and for the importance of adopting common standards if the home VCR—*anyone's* home VCR—was to gain acceptance. Two weeks later, Tetsujiro Nakao, the executive vice-president of Matsushita, telephoned Kazuo Iwama, the new president of Sony (and Morita's brother-in-law), to say, "Please wait three months longer on your proposal."

Not liking the sound of that, Morita took an extreme step. He went to see Konosuke Matsushita, with whom he had never talked business before. Again he made a plea for uniformity. "Price and compatibility will be the deciding factors in spreading a big new product like the home videocassette recorder throughout the market," he said. "Price goes without saying, but if different companies are divided and their tapes are incompatible, the 'soft' part of the business will not progress." He was thinking of prerecorded videocassettes, a market Sony hoped to encourage when VCR sales reached a suitable level. "That will put a brake on the spread of the hardware," he added.

Konosuke Matsushita was vague about his company's intentions—in part, perhaps, because he wasn't sure of them. But he spoke of the VX-2000 and of JVC's work—not hitherto disclosed—on a format called the video home system, or VHS. Maybe it would be useful, he suggested, for the top managements of all three companies to meet and compare their machines.

The meeting took place at JVC's Tokyo offices on April 5, just as the cherry blossoms had begun to open. From Sony, Morita, Iwama, and Deputy President Norio Ohga attended; from Matsushita, Konosuke Matsushita came with Executive Vice-President Takayoshi Inai, who, in his other capacity as president of Matsushita Kotobuki, had overseen the development of the VX-100 and VX-2000. From JVC, Matsuno, the president, came with Tokumitsu, his executive vice-president. In the room were samples of the three companies' VCR's, and the first item on the agenda was the simultaneous playing, through all three, of a "Sesame Street" type of children's program called "Ping Pong Pang," which had been recorded that morning. Everyone watched raptly, although the Sony contingent's eyes were on the JVC machine as much as on the picture it produced. When the performance ended, Matsuno and Tokumitsu invited their guests to look inside their prototype, and Morita, Iwama, and Ohga took them up on their offer.

What they saw was a considerably smaller machine than the Betamax (although it was hard to judge the significance of that since prototypes are often smaller than production models). Mechanically, too, VHS had a notable distinction: the use of a loading system called M-loading, in which two small posts instead of one big one pull the tape from the cassette and move it directly to the drum instead of taking the circuitous, back-door route of the U-loading system in the Betamax (and the U-matic). M-loading was not unfamiliar to the Sony contingent. The basic concept had been tried in some of the early U-matic prototypes, and the company had filed a patent on it in 1969. In other respects, JVC'S and Sony's machines were strikingly similar. Both were two-head, helical-scanning machines using half-inch tape in a U-matic type of cassette. Both—unlike the V-Cord, the VX, and indeed all the color video recorders to date—used azimuth recording and countered the problem of cross talk by juggling the phase of the color signal. So the Betamax and the VHS were in a class by themselves as far as tape efficiency went. The real difference between them lay in *how* the two companies had chosen to exploit that advantage: Sony to make the cassette paperback size, and JVC to achieve a two-hour recording capacity (with a cassette that was about a third larger).

Patent infringement was not at issue since Sony, Matsushita, and JVC still had a free-exchange agreement on their video-recording patents. Even so, JVC's behavior looked decidedly underhanded to the men from Sony, and they could not easily find words—words appropriate to a Japanese business meeting—for what they felt.

Eventually one of them suggested that the two machines were basically the same except for the larger cassette size of VHS, and he asked: "Isn't the VHS changed only to record two hours?"

Now it was Matsuno's and Tokumitsu's turn to get testy. They agreed that two-hour recording was important—indeed, vital—but so, they said, were the other differences, notably M-loading and a JVC-developed recording process known as dual-limit frequency modulation, or DL FM.

At last Morita said what all of the Sony representatives were thinking: "It's a copy of Betamax." And Iwama and Ohga—"with fists clenched," according to one account—added words to the same effect.

The JVC people had prepared a kind of script for this meeting. Matsuno and Tokumitsu were to conclude by urging Sony to adopt VHS, in the interests of uniformity, and Konosuke Matsushita was to lend his endorsement. But the script, and all pretense of diplomacy, went out the window as soon as the word *copy* had been spoken. Tokumitsu, Morita's old naval superior, responded vehemently. VHS, he said, was the product of an R&D program that went back to a time well before anyone at JVC had seen the Betamax. "Two hours will be the mainstream for home video, so there is a basic inadequacy to the Betamax," he said. "Technology is moving forward by leaps and bounds, so it is natural that the product coming out later is better."

In weeks, months, and years to come, Morita and his colleagues would wonder why, if JVC had been working on its system for so long, there had been no mention of it on the occasion of Sony's initial offer in the autumn of 1974. Was it because the system would have been in too primitive a state to show? Was it possibly the case that JVC's engineers had needed the opportunity to examine a Betamax prototype before they had been able to overcome the central problem of reconciling azimuth recording with a satisfactory color picture? One way or another, it seemed to Sony, JVC had not played fair. "With U-matic, we were a family, and in a family you should not be so secretive," Morita said later. "If they had told us what they were doing, we could have come to a reasonable arrangement."

When these queries were put to some of the top people at JVC, they responded that it would have been pointless to ask Sony to adopt their system because Sony would not have listened. Sony, they said, was simply too confident—a euphemism, perhaps, for *arrogant.* In any case, in the JVC view of things, Sony had no right to complain about unfamilylike behavior, since (as it appeared from an item in one of the trade papers) it had shown a Betamax prototype to Anthony Conrad, the president of RCA—a *foreign* company no less—before it had shown one to any of its fellow family members. And the argument would spiral on from there.

Not all these points got a full airing at the April 1976 meeting, however, for Morita still hoped to persuade Matsushita, if not JVC, of the Betamax's superiority, and he had an ace up his sleeve or, to be more exact, in the garage. At a signal from Morita, Iwama and Ohga left the room and, a few minutes later, returned with a prototype of a Betamax capable of recording for two hours, with only a marginal sacrifice of picture quality. In Morita's mind, this latest achievement by Sony demolished JVC's whole argument. But Tokumitsu and Matsuno persisted in arguing that recording time was only one of the departments in which VHS was superior. M-loading, they said was mechanically simpler, the machine itself lighter and inherently easier to manufacture.

The VX-2000, Matsushita Kotobuki's contribution to the world of VCRs, was ignored throughout the discussion, to the obvious displeasure of Takayoshi Inai, who had presided over its development. The air was so crowded with the bad feeling between JVC and Sony, however, that Inai could not bring himself to say much. Konosuke Matsushita also fell into a listening attitude, and when the meeting broke up, Morita invited him to visit Sony's R&D lab and its VCR plant in Aichi, in the hope that the advanced state of Sony's research and production facilities would make JVC's system look like a flight of fantasy by comparison. Matsushita accepted on his own behalf, but when he asked his JVC colleagues to join him, Tokumitsu sternly refused, explaining, "It's not right to see the laboratories of a rival company making incompatible products."

"We have done all we can" was Morita's assessment when, later in the day, he finished showing Konosuke Matsushita and Inai through the Sony research lab and saw them off.

Three weeks later Morita received an invitation to meet with Konosuke Matsushita. This time they met alone, at Matsushita's Tokyo offices, with subordinates from all three companies waiting anxiously outside. Konosuke Matsushita began with an apology. It was too bad, he said, that he had not been made aware of Sony's standardization proposal from the beginning. Had he known about it then, "something might have been worked out." It would be better for all three companies, he agreed, to adopt a common standard. But with the VX-2000 already in mass production and JVC firmly committed to VHS, it was too late for that. He took the opportunity to add that he personally regarded VHS as superior to Betamax. To Morita's consternation, Konosuke Matsushita had bought JVC's claim of greater simplicity. "As I see it," he said, "if you give the Betamax a hundred points, then the VHS merits a hundred fifty. If the VHS had been a hundred twenty points, I could have pushed through the Betamax format, but . . . Since the difference is so great, I want Sony to use the VHS."

When these words reached Nobutoshi Kihara, they burned. He and his fellow engineers fervently believed that the Betamax was the better machine. M-loading, in their opinion, wasn't really simpler than U-loading, although it might seem so to a nontechnical person who examined the JVC and Sony machines in a casual way. The loading mechanism of the Betamax looked complicated because it pulled twice as much tape free of the cassette and carried it around to the far side of the drum instead of the near side. But the mechanical apparatus that performed this feat was ingeniously simple. Besides, the elaborate route followed by the tape in a U-loading machine served a purpose;

it was a way of reducing stress, especially at the high speeds of the rewind and fast-forward modes. In M-loading there was not only less tape to absorb the stress but more stress to absorb because the tape had to go through two sharp turns (the tops of the M) instead of one. (As a result, the tape in the early VHS machines had to be withdrawn into the cassette before rewinding or fast-forwarding, which ruled out a "picture search" mode.)

A soft-spoken man who measures his words, Kihara would not go so far as to say that JVC's system had been copied from Betamax. But he noted that both had a common ancestor: the U-matic. "When we made the U-matic, we showed them our entire hand," Kihara said, "and the discussions were settled on the idea of mutually cross-licensing the patents. It was all done in the name of the noble cause of standardization." He paused and added, "It may have been a mistake from the start."

RAISING A FAMILY

LOOKING BACK ON their year-and-a-half-long courtship of Matsushita and JVC, Morita and his colleagues could not help feeling that they had been led on. All those politely ambiguous responses, the requests for more data, the vague reassurances, the pleas for patience—what had been behind them? One thing that had, it later became apparent, was a secret campaign by JVC, beginning in the summer of 1975, to form an alliance of VHS companies. No one would come right out and say that Sony had been the victim of a cunning plot by JVC to gain time and support for its own product. But if that had not been the *purpose* of JVC's actions, it had certainly been the *result.*

JVC's executives had known all along that they were in a battle with Sony, while Sony's executives had not quite known *what* was happening. In their bewilderment they had made at least one decision they would come to regret. In July 1975, nine months before the fateful meeting in JVC's boardroom, the management of Hitachi had approached Sony to ask for a license to manufacture VCRs in the Beta format. Sony had reluctantly rejected that idea, largely out of concern over how it would be taken by Matsushita (Matsushita and Hitachi being the Macy's and Gimbels of the Japanese appliance industry). Unaware of JVC's plans and fancying that Sony had a special relationship with Matsushita, Morita and his associates had thought it best to settle matters with Matsushita first and then and only then seek out other companies. In turning down Hitachi, Sony had tried to hold out hope for an arrangement between the two companies at a later date. But Hitachi had taken the rebuff to mean that Sony intended to keep the Betamax technology to itself, an assumption that had made Hitachi ripe for the advances of JVC, whose "VHS family" it proceeded to join.

In the spring of 1976 Morita was faced with the realization that Hitachi and Matsushita, as well as Sharp, Mitsubishi, Toshiba, and Sanyo (the last two companies having come more or less to the end of the road with their V-Cord system), were preparing to join the VHS family. As the journalist Masaaki Satoh put it, "Sony stood at the edge of the Inland Sea."

In its desperation Sony looked to the Japanese government for help. The proliferation of home VCR formats was already a subject of concern at MITI, which feared (as Sony had been arguing in vain to JVC and Matsushita) that Japan's entire electronics industry would pay for the failure to get its act together. Now, it seemed, JVC was about to introduce yet another format. Sony's word carried great weight with MITI because of its excellent record as an exporter, and when Morita and Iwama argued that another format would confuse the public and hurt sales for everybody, MITI was inclined to agree; it was also inclined to agree that the Betamax—as the first successful home VCR to reach the market—was the logical standard for the industry to unite behind. When JVC heard that argument, however, it responded with a "numbers theory"—the industry should go with the format favored by the most companies—and it bombarded MITI with evidence of the superiority of *its* system.

MITI took the issue seriously enough to initiate a study and to ask JVC to defer its production plans. The basic strategy of MITI officials, however, was to gather representatives of all the companies in a room, plead with them to adopt a standard, and wait for someone—anyone—to back down. Toshiba and Sanyo were good enough to say that if a consensus developed, they would join it. Matsushita, for its part, volunteered to abandon the VX format (which the public was already abandoning) in favor of any reasonable alternative with more than a one-hour recording capacity. Neither Sony nor JVC was prepared to consider any format but its own.

On August 11, Kokichi Matsuno, the president of JVC, was summoned to a meeting with MITI officials and urged to renounce VHS and adopt the two-hour version of the Betamax, which Sony would soon be ready to put into production. If he went along, they promised, MITI would force Sony to swallow JVC's development costs. To Matsuno this was a contemptible proposal, but the word *no* rarely falls from a Japanese businessman's lips. The question, he politely explained, was not for JVC to decide alone. He would relay MITI's recommendation to the members of the VHS family and solicit their views. A month later JVC delivered its answer in the form of a public announcement: VHS would go into production in October.

It was MITI, in the end, that backed down. In October a socialist legislator raised the VCR compatibility question on the floor of the Diet, as a case study in the failure of private enterprise to join forces in the national interest. What was MITI going to do about it? A MITI spokesman explained that it was probably too late to do anything about half-inch video, but he feebly added that the electronics companies were under orders to negotiate a standard for the *next* generation of VCRs.

Even before its big-government strategy had played itself out, Sony began probing for cracks in JVC's family. Suspecting, correctly, that firm agreements had yet to be signed in all cases, Sony's management divided up the task of visiting the companies involved. This time Sony's ambassadors went to great lengths to show their sensitivity to each company's concerns. Morita, in particular, impressed the people he spoke to with his readiness to apologize for past stubbornness and overconfidence; he had been wrong, he said, to have insisted that a one-hour system was enough. "I had heard rumors about Chairman Morita's ego," Hirokichi Yoshiyama, the president of Hitachi, said later, "but he was more stronghearted than I expected."

Hitachi had already made a firm commitment to VHS, and Sharp and Mitsubishi also proved to be loyal members of the VHS family. But in Toshiba and Sanyo, Sony found a pair of potential runaways. Both had advised JVC to make its cassette smaller and to reconsider the use of the M-loading system, which, in their judgment, put more strain on the tape than Sony's U-loading. When JVC failed to follow their advice, and when Sony came along with its new two-hour model and its new attitude of penitence, they listened closely. It probably didn't hurt that Morita and Toshiba's president, Kazuo Iwata, both were sons of sake brewers from Nagoya. Morita made his initial overture in July, and he and Iwata played golf together often that summer.

JVC, meanwhile, was having trouble getting its family together for a grand announcement. Hitachi, Mitsubishi, and Sharp had to negotiate the rights to use certain video patents owned by Sony, and until those negotiations were resolved, they weren't going to do anything that Sony might find offensive. As for Matsushita, it spent months studying JVC's system from a manufacturing standpoint but appeared to be in no hurry to make a formal commitment, largely, it seemed, because of the need to unload an accumulated backlog of VX-2000s—a format that threatened to become extremely unattractive if retailers and consumers knew that Matsushita was preparing to embrace VHS. Nevertheless, by the beginning of 1977 Hitachi, Sharp, and Mitsubishi had committed themselves, and on January 10 Matsushita declared that it would put a VHS model into production in the spring, although the company expressed continued support for the VX-2000 as a "low-priced alternative."

A month later Sony unveiled the results of *its* recruitment efforts. Not only had Toshiba and Sanyo agreed to join the Beta family as manufacturers—as the industry expected by now—but Zenith, the American TV manufacturer, had agreed to buy Sony-made VCRs on an OEM basis. For Sony, this meant forsaking an old company policy: never to sell under any brand but its own. But the stakes were more important than the principle.

The news about Zenith came as a shock to JVC and Matsushita. Sony already had a commanding sales advantage in the United States, where JVC (the only company that actually had VHS machines in production) was a small influence. JVC's top people could not help remembering Matsushita's laggardly response to their original invitation. If Matsushita had declared itself the previous summer, they pointed out, Toshiba and Sanyo would probably not have had the nerve to go with Betamax, and Zenith, too, might have thought twice about adopting an also-ran format.

At Matsushita the Zenith revelation sparked a more constructive what-if: What if Matsushita could make an OEM deal with RCA? The new president of RCA, Edgar Griffiths, had recently issued a disdainful statement about the home VCR market, which, he had said, RCA had no intention of entering in 1977. RCA itself had been unable to make a satisfactory VCR at a reasonable price, and it had shifted its corporate attention to a form of videodisc player. Officially RCA's management took the position that magnetic tape was too expensive for a consumer product, and the videodisc venture certainly gave RCA a reason to hope this was true. On the other hand, RCA's disc system was still a few years away from the market, and surely (the management of Matsushita reasoned) RCA would not want to give Zenith, its major American competitor as a TV maker, a clear path in the VCR field.

A trip to the United States was in order, and Masharu Matsushita, soon to be elevated from president to chairman, decide to make it. He was joined by the general manager of the company's planning and engineering division, Hiroshi Sugaya, and its executive vice-president, Takayoshi Inai, who was intimately familiar with the problems of manufacturing VCRs, having supervised Matsushita's own contributions to the VCR sweepstakes, the VX-100 and VX-2000.

In Edgar Griffiths, Matsushita had the good fortune to encounter an American business leader who was not ashamed to put his company's name on products made by others. RCA's real strength lay in marketing power, Griffiths reasoned so why should it have to prove its machismo by manufacturing things? "Buying when it makes sense to, manufacturing when it makes sense to" was the new RCA policy enunciated by Griffiths, and despite his public pronouncements on the VCR question, he had privately talked about possible OEM deals with Sony (which had named a price that RCA deemed too high) and with JVC (which had proposed a delivery schedule that RCA deemed too protracted). So the Matsushita delegation had reason to be optimistic. "Knowing RCA's conclusion that a two-hour recording time was vitally necessary, we thought the VHS system would be just what they wanted," Tsuzo Murase recalled. "But the very day of the departure of our representatives for the United States, Sony announced that it would be selling the two-hour version of Betamax." With Sony in the two-hour business, one of the major advantages of VHS—possibly *the* major advantage—had evaporated.

In New York the three Matsushita executives were joined by Akira Harada, the president of the Matsushita Electric Corporation of America (MECA), and they met with a group of RCA executives led by Griffiths and Roy Pollack, vice-president for consumer electronics. The visitors had brought a prototype of the VHS machine Matsushita was getting ready to produce. After watching it in action, the RCA people pronounced themselves pleased with its performance and prepared to talk about an OEM contract. But first they had a small modification to propose. The two-hour recording capacity, one of them said, "lacked sex appeal." RCA wanted a machine that could record a football game. That meant a recording time of at least three hours. If Matsushita could work its way up to that neighborhood, Griffiths and Pollack suggested, RCA would be ready to do business.

Matsushita had only just succeeded in designing its two-hour model, and the notion of trying to make such a significant change so soon was, at first, bewildering. But Sugaya had observed the tendency of American businessmen to become abruptly unavailable when football games went on the air. Maybe two hours *did* lack sex appeal, he said to himself. And Inai, who only a few months earlier had passionately defended the VX-2000 and urged his company not to adopt VHS at all, thought back to April 1976, when he had seen Sony's two-hour prototype. Sony had turned a one-hour machine into a two-hour machine by managing somehow to halve the track pitch—the width of each track— and the running speed. The track pitch and running speed in VHS were almost the same as in the original Betamax. What Sony could do, Matsushita could do.

Griffiths and Pollack were asked how they would feel about *four hours.* They said they would feel fine.

In Japan Murase was busy overseeing the start of production on Matsushita's two-hour machine. When Harada called to explain what RCA wanted, Murase's first reaction was: "Impossible."

"It's not a matter of possible or impossible," Harada replied. "You must."

The company set up a project team with Murase as design strategist,

Inai in charge of preparing the production lines (a TV plant on Shikoku Island would be turned into a VCR plant), and Akio Tannii, the head of the video division, doing the coordinating between the lab and the factory. Every available hand and every available resource was enlisted. Seventy engineers worked on the project over the next several months, and some worked twice their usual shifts. The video division had said it would need six months merely to construct a prototype. But RCA wanted to have machines to *sell* in time for the Christmas buying season.

One resource the Matsushita engineers did not succeed in tapping was their colleagues at JVC—the people who had developed VHS in the first place. Asked for advice on going from two hours to four, JVC advised Matsushita to drop the idea. Tokumitsu pointed out that JVC had experimented with a four-hour version, only to decide that it entailed an unacceptable loss of quality. JVC was criticizing Beta II, the two-hour Beta format, as a bastardization of Beta I, and now it was being asked to condone the same kind of compromise in its own system. The JVC people were also concerned about the issue of compatibility; although it would be possible to play two-hour recordings made on JVC's machines through Matsushita's, it would be impossible to play four-hour recordings made on Matsushita's through JVC's. (In consumer electronics parlance, there would be "upward compatibility" but not "downward.") The JVC people had the feeling that the parent company had seized control of their design. "We lent the eaves and the house was taken," one JVC official complained.

It turned out to be a lucky thing for Matsushita that production had already begun on its two-hour machines (which wound up being sold exclusively in Japan); for they became the prototypes on which the engineers could try out various ideas as they struggled to accomplish the four-hour project. The job that had looked as if it would take six months wound up taking six weeks. On March 22 and 23 Matsushita held parties in Osaka and Tokyo to celebrate Toshihiko Yamashita's installation as president. Although only a few top people in the company knew it, there was more to celebrate. The four-hour prototype was finished, and the finishing touches were being put on a contract with RCA.

A week later, on March 30, Yamashita met with Pollack in Osaka to exchange signatures on a document that obliged Matsushita to make—and RCA to buy—some fifty-five thousand machines that year, and between five hundred thousand and a million machines over the next three years. The entire world's production of home VCRs to date was in the low hundred thousands. RCA was either going to wind up with an awful lot of VCRs in its warehouses or going to give the business a jolt.

For RCA the agreement was the end of a long trial. Twenty-six years earlier David Sarnoff had spoken of his desire to see RCA produce "a television picture recorder that would record the video signals of television on an inexpensive tape, just as music and speech are now recorded on a photographic disk or tape." If Sarnoff were still alive and could see the Matsushita machine, the RCA engineer George Brown observed, he would surely have said, "It took a little longer than it should have, but the result is exactly what I had in mind."

In the summer of 1977, at the annual convention of the International Tape Association on Hilton Head Island, South Carolina, Morton Fink, an executive of the consumer products division of Sonam, visited the Panasonic suite and took a look—his first—at one of the Matsushita-made VCRs that had begun to trickle in from Japan. The Panasonic salespeople were talking up several features that could be found on their machines and not on Sony's: a built-in clock, a remote pause control, and, above all, four-hour recording. But Fink was more impressed by the basic construction of Matsushita's product. Although sleeker-looking than the Betamax, Matsushita's VHS machine was about half again as large as JVC's, leaving more space between components. Unlike every other VCR to date, it had a direct-drive motor rather than a belt-driven one. And it had an all-aluminum, all die-cast chassis, instead of the hybrid structure—partly die cast, partly pressed metal—of the Betamax. These things tended to reduce the cost of labor and materials at high volumes of production. Matsushita had made a product that was combat-ready for a price war.

"At Sonam," Fink said, "we always had Japanese colleagues in various jobs—mainly the product jobs. They were the link with the factories, and they had a little Falcon jet at their disposal. They must have flown three engineers down the next day to go in there and take a look at that chassis and see that it was really die-cast."

Sony went into even more of a tizzy in August, when RCA announced that its Matsushita-made VCRs would sell for three hundred dollars less than the "current leading competitive VCR instrument"—otherwise known as Betamax—and would go on sale in September, only six months after the contract with Matsushita had been signed. With each machine RCA would offer a package of free prerecorded tapes (including Muhammad Ali fight films), and it would pump some four million dollars into an advertising campaign. The message of the ads was simple: "Four hours. $1,000. SelectaVision." (SelectaVision was the name RCA had been clinging to for nearly a decade, while awaiting a product worthy of it.) It had been two decades since RCA had brought out a significant new consumer product, but its distribution and marketing machinery remained formidable. "We really orchestrated that distribution," William Hittinger, RCA's executive vice-president for consumer electronics, said later. "We broke it very hard and very rapidly. It was a blockbuster promotion. We spent four months working on it."

RCA's price surprised even Matsushita. "At that price RCA not only couldn't make a profit but should have lost money," Matsushita's president, Toshihiko Yamashita, noted disapprovingly. Industry analysts estimated that SelectaVision had an FOB (free on board) cost of about $500, and duties, domestic shipping, and marketing expenses probably brought that sum up into the neighborhood of $900. Whether or not RCA could make money selling VCRs at $1,000, Matsushita's surprise was apparently genuine. The Japanese counterpart of RCA's machine, a two-hour model known as the MacLord (after the Japanese transliteration of the American TV detective series "McCloud"), was selling for 267,000 yen, or almost the same as RCA's proposed price, and as the yen rose in value that summer and fall, MacLords became more expensive in Japan than SelectaVisions were in the United States. The people who ran Matsushita's American subsidiary, MECA, had intimations of second-class citizenhood. They intended to sell the same basic machine as RCA's under the Panasonic brand, for a recommended list price of $1,280. Now they hurriedly reset their price to $1,095.

The Betamax, at $1,300, was out on a limb, but Sony proudly declared that it would not be drawn into a price war. "If RCA sells its video recorders cheaply, it will stimulate demand, so in the long run it will be a plus to Sony as well," Morita explained. "Businessmen are in a kind of marathon. The runner who sets his own pace will hold on to win. We've just passed the first milepost in the VTR war." Sony had "never once been influenced by the pricing strategy of another company."

But there was always a first time. Zenith had no intention of being undercut by RCA. On November 4 it dropped the price of its Betamax clone from $1,300 to $994, thus acquiring the right to claim the least expensive VCR on the market. Sony could hardly let itself be underpriced this drastically by an OEM client. Ten days later Sony lowered the basic Betamax price to $1,095.

As humiliating as it must have been for Morita to reverse himself so quickly, worse humiliations were in store. By March 1978 Americans were buying more SelectaVisions than Betamaxes. "The longer playing time turned out to be very important, and RCA's product was better styled," Robert Brown, the vice-president of marketing for Zenith, recalled. RCA, which did nothing but "buy, sell, and advertise" its VCRs, as an article in *Fortune* magazine put it, had a 28 percent market share for the first quarter of that year, while Sony had a 27.5 percent share. For 1978 as a whole, RCA's share was 36 percent, and Sony's 19.1.

Matsushita's triumph as a manufacturer was even more dramatic. After the RCA agreement had come a series of OEM contracts, each adding to Matsushita's economies of scale and putting its would-be competitors at a greater disadvantage. By early 1978 an American consumer could walk into a store, choose any of seven brands of VCR—General Electric, Magnavox, Curtis Mathes, Montgomery Ward, Sylvania, RCA, or Panasonic—and walk away with essentially the same Matsushita-made product. Before long Matsushita was turning out two-thirds of the world's home VCRs. And with JVC and Hitachi also exporting VHS machines to the United States, the Beta family began to look woefully small.

By going to two hours and cutting prices, Sony had not arrested the trend away from Betamax, but it had brought on a serious customer relations problem. Suddenly the original Betamax customers—the first-on-the-blockers who had paid thirteen hundred dollars or so for a one-hour (Beta I) machine—were angry. They did not enjoy seeing their neighbors come home with two- and four-hour machines which cost significantly less money. They did not enjoy learning, in the fall of 1977, that Sony planned to eliminate the one-hour speed from some of its two-hour (Beta II) models or finding out, around the same time, that the prerecorded Beta videocassettes that were beginning to appear on the market were only for the two-hour models. It looked as if *their* Betamaxes were headed for early obsolescence (although, of course, they could still do everything Sony had claimed they would do). Sony had vowed to come up with a "cassette auto-changer" to allow for two-hour recording on a one-hour machine, but the changer was nowhere to be seen. A few angry customers went so far as to file lawsuits, accusing Sony of breach of something or other, and though a Sony spokesman protested that "customers don't sue General Motors when they introduce a new model," the company was obviously stung by the criticism.

The auto-changer finally surfaced at the beginning of 1979, and it was offered to Beta I buyers at a special "customer satisfaction" price of fifty dollars. Sony had probably never come up with a less inspiring product. A reviewer described it in the pages of *The Videophile* as "a complicated mother, no doubt about it. Upon examining it carefully, one gets the impression that some poor demented Japanese engineer spent six months of his life working on the device until his mind cracked like a piece of bent chalk. As a matter of fact, we understand that Sony has a special loony farm just a few miles from their factory . . . for engineers who have either gone nuts working on the changer or simply had breakdowns in trying to combat VHS."

As a rule the video magazines gave Sony's products high marks. People who prized technical excellence above everything else tended to champion the superiority of Beta over VHS. "In my opinion, the Sony-developed Beta format is superior to VHS in several ways, including better cassette design, superior tape handling, and overall better video engineering . . ." a columnist wrote in the December 1980 issue of

Video. He praised the U-loading system for the ability to fast-forward and reverse without disengaging the tape from the head. "Leaving the tape wrapped during all operating modes reduces waiting time between mode changes. On the Beta 5600 and 5800 (and some more recent 5400s) you can even go directly from one mode to another without pressing stop. . . . The M-load system repeatedly loads and unloads the tape between operating cycles. This action is accompanied by mechanical noise and a time delay." But such words were cold comfort to Sony while its market share continued to shrink.

The auto-changer was only the least and crudest of a series of innovations with which Sony tried to shore up the Betamax's popularity. A new cassette with thinner tape increased the maximum recording time to three hours. Yet another adjustment in track pitch and running speed, followed by the introduction of a cassette with even thinner tape, raised that figure to five hours. BetaScan, Betamovie, and Beta HiFi came along, and with each refinement Sony drew praise from reviewers and a spurt of fresh interest from consumers. But nothing made more than a passing difference, and contrary to Sony's assertions, JVC and the other VHS manufacturers found a way, in time, to respond to every maneuver. Once it became clear that more people were buying VHS than Beta, not only in the United States but also in Japan and Europe, compatibility became VHS's ally and Sony's enemy. By the middle of 1979 VHS models were outselling Beta by two to one in the U.S. market. Even when the price of no-frills Beta models dropped sharply below the price of VHSs, even when Beta machines acquired the ability to record for as long as any reasonable human being could wish, Sony went on paying the penalty for its original sin. And even then few experts would disagree with Masaaki Satoh's judgment that "if Matsushita had developed Betamax, it would probably be the standard format in the world today."

WILBUR AND ORVILLE AND TOM, DICK, AND HARRY

TO GET FROM downtown Hollywood to the birthplace of the prerecorded videocassette business, you head east on the Pomona Freeway for about two thousand miles, turn left, and proceed another thousand miles or so until you come to Farmington Hills, Michigan. Farmington Hills is a suburb of Detroit, and in the fall of 1976 a person had to be about that far from Hollywood to believe that the VCR and the movie industry could be something other than mortal enemies—and to be ready to put his money where his mouth was.

Andre Blay had spent most of his life in Michigan. At thirty-nine he had already forged a successful business career as a distributor and servicer of audio and video equipment and, on a small scale, a producer and duplicator of videotaped material, mostly in the U-matic format, for advertising agencies, corporate training departments, and the like. But the entrepreneurial hormones were still flowing, and they needed a new outlet. For several years Blay had been thinking about prerecorded video as a field of opportunity. When the Betamax system graduated to two hours and JVC began exporting its two-hour VHS machines to the United States, Blay decided the time was right. So he wrote a "cold call letter," as he described it, inquiring about the right to sell movies on videocassettes, and he sent it to the chief executive of every studio in Hollywood—every studio, that is, except Universal, which, he gathered from reading the papers, would not be amused.

Only two studios bothered to respond to Blay's letter, and one of them, MGM, simply informed him that the matter—not his proposal in particular but the relationship of the movie industry and the VCR in general—was under study. Movie executives, as a class, do not get too excited about letters from people they have never heard of who run companies they have never heard of in places they have never heard of. Even if Blay's inquiry had received the full time and attention of the people to whom he addressed it, the response might not have been much different, for his idea went against a tradition as old as Hollywood itself: that of never parting with title to a copy of a motion picture. Blay was asking the industry to go back to a way of doing business that it had deliberately put an end to sixty years earlier.

At the beginning of the century producers were in the habit of selling their movies outright—either directly to exhibitors or to local "exchanges" which supplied the exhibitors in a particular part of the country. In 1908 Thomas Edison and a group of rival inventors and entrepreneurs settled a long legal brouhaha over patent rights on movie cameras, projectors, and film by joining forces under the banner of the Motion Picture Patents Company. This body, which came to be known as the Trust, claimed absolute control over the production, distributions and exhibition of motion pictures, and to help make the claim stick, its member companies adopted a policy of retaining title to every print and selling the *right to exhibit* a movie rather than the movie itself. This allowed them to shift prints around the country over a period of months in order to reach a rapidly expanding audience. It also meant that the studios, through their rental contracts, could impose a fee structure based on the size and earning power of each theater and could have a large say in such matters as the length of engagements and the price of tickets.

With only ten charter members and no provision for adding to them, the Trust was too clubby for its own good. Even if its legal position had been ironclad, it could never have satisfied what turned out to be a ferocious national appetite for movies, and it could never have afforded all the litigation that would have been necessary to beat back the countless competitors who were drawn into the business. By 1915, when a federal court ruled the Trust to be a trust in the legal sense—a monopoly—it was already collapsing. But in its brief reign it had played a powerful part in imposing a new, vertically integrated order on the business, and the "independent" movie companies that arose in opposition to the Trust adopted very similar marketing policies and gained a similarly tight hold on distribution and exhibition (or, in some cases, began as distributors or exhibitors and achieved vertical integration from below). Because of other changes in which the Trust had less of a role—the rise of the star system, the big screen, and the long narrative—moviemaking was an exceedingly expensive and complicated business by the beginning of World War I. Sheer momentum allowed a handful of studios to dominate for decades, and so, in a sense, the demise of the Trust might be termed a pyrrhic defeat—not for its member companies, which *were* defeated, but for oligarchy in the abstract, which endured.

The policy of owning every print served the industry as valuable legal armor when another antitrust prosecution, in the 1940s, compelled the studios to divest themselves of their theater chains. Rental is by nature a more intimate business arrangement than sale. The tie between seller and buyer ends with the transaction, and the buyer is free to pass his property on to someone else. With rental, the transaction is just the beginning of the relationship, and the lessor, if market conditions are right, can dictate elaborate instructions to the lessee about how he must conduct himself.

A lease arrangement also made it easier for the studios to deal with movie pirates. If a studio could go into a court of law and flatly state that it had never parted with title to a copy of any of its films, the accused pirate could not easily mount a defense based on the claim that he had, say, bought the movie from a stranger on a street corner. The coming of the home VCR did not immediately strike most studio executives as a reason to abandon this policy. On the contrary, the VCR had set off a wave of piracy worse than any the studios had seen before. In parts of the world where standards of public behavior, censorship laws, or trade protection policies tended to inhibit the exhibition of American movies, VCRs turned homes into movie palaces, and an elaborate network of illegal distribution came into being with remarkable speed. It was like the traffic in certain illegal drugs, only reversed. America was the source of supply, and the contraband flowed out along a convoluted route through a series of middlemen. As soon as a new movie opened, a print here and a print there would disappear (perhaps so briefly that no one noticed), and before long videocassettes of it would show up in Saudi Arabia, South Africa, and Singapore. It was hard for Hollywood to exercise control over events in such remote places—not great bastions of copyright protection, in any case. The best hope for managing the problem seemed to lie in cutting off the flow at the source. A pirate had to go to some trouble, the industry could console itself, to steal a thirty-five-millimeter print of a movie and arrange for its transfer to videotape. If movies were freely and legally available on videocassettes, piracy would become as easy as hooking up two VCRs.

Few studio executives felt as passionately about these matters as Sidney Sheinberg, but the general attitude toward the VCR was the same as his: fear. If there was to be some kind of home market for movies, the studio people, by and large, hoped it would develop through one of the videodisc systems which were about to come on the market—systems that denied the customer the power to record and thus to copy.

But the 1970s were a tumultuous time in the history of Hollywood. The old studio patriarchs had died or been dispossessed, and with them had gone a deep reservoir of institutional memory. Several of the studios had been bought by conglomerates which were out to prove that Hollywood had something to gain from the insights associated with a "broader" business perspective. The new Hollywood was increasingly run by lawyers, agents, and M.B.A.'s—people who could navigate more smoothly in other regions of the business and financial world. The Hollywood tradition that made the biggest impression on these people was the fact that lately a lot of movie executives hadn't lasted very long in their jobs. Power was being doled out on a shorter- and shorter-term basis. If the studio made a nice profit this year, the management could expect to be around next year. If not, not.

In 1976 Twentieth Century-Fox had been through a few difficult years, and there was not a lot of idle cash sitting around in its money bin (although there would be plenty a year later, thanks to *Star Wars*). So the company was ripe for a new idea, especially an idea that didn't cost much money up front and promised to make some in a foreseeable accounting period. In addition, Fox had a subsidiary known as Twentieth Century-Fox Telecommunications and, running it, a young executive named Steven Roberts, who bore a resemblance to Dick Clark of "American Bandstand" fame. Cable TV was Roberts's principal domain, and it seemed to him that Hollywood had misjudged cable at first, seeing only a threat to movie theaters rather than a way to reach a new audience. Roberts took it to be part of his mission to save Fox from making such a mistake again. "I took the position that I never wanted to see our company let a new technology come along and bury its head in the sand," he said later. It behooved Fox, Roberts felt, to find out just how much demand existed for movies on videocassettes and if there was some way to satisfy it without giving aid and comfort to the pirates. Maybe the way to beat them was to compete with them.

Andre Blay's letter found its way to Roberts just as he was preparing to run an experiment. "We were about to go into the business ourselves—to test the waters," Roberts said. "We would use films that were already on commercial television, so I felt that we were risking very little, since anyone who wanted them could tape them off the air as it was. But now we were going to sell those films without the commercials, and unedited, because on commercial television they were edited to fit the time slots." For a trial venture it seemed sensible to let somebody else invest the capital, and here was Blay, ready, willing, and apparently able. Still, it took them the better part of a year to work out a mutually satisfactory arrangement and to overcome the piracy fears of other Fox executives. The contract they signed in July 1977 called for Blay to pay Fox an advance of $300,000, plus a minimum of $500,000 a year against a royalty of $7.50 on each cassette sold. He would have the nonexclusive right to choose fifty titles from a list of a hundred prepared by Fox—all of them movies previously sold to network TV, none of them released more recently than 1973.

Blay's company, Magnetic Video, already had a few videocassette duplication machines. But the new venture was on a different scale.

"Until then fifty copies a week was a lot of copies," Blay said. "No one was thinking of consumer volumes. There was no one equipped in the world to undertake this mission. So we had to take the gamble. In addition to investing in the film rights and the marketing program, we had to build our own factory—a facility to make twenty thousand Beta and VHS cassettes a month. That was an unheard-of figure at the time, although now there are factories that make half a million or more."

It cost him about one and a half million dollars to launch his business. He borrowed most of the money from a bank, which questioned the wisdom of his project but—in a view of the credit he had built up with his existing operations—"had no legitimate reason to deny me use of funds." From the movies offered by Fox, Blay based his choices on *Variety*'s estimates of their box-office receipts. He selected *Hello, Dolly!, M*A*S*H, Patton, The French Connection, The King and I, Beneath the Planet of the Apes, The Sound of Music, Butch Cassidy and the Sundance Kid,* and forty-two others.

By October Blay was ready to start selling, although not quite sure how to go about it. He began by taking out ads in the trade publications that catered to record stores and "brown goods," or appliance, stores. "Dealer inquiries invited," the ads said. The brown goods stores were "ill-equipped and lacked incentive to build a long-term business," Blay said later, but he saw them as "a bridge" to carry him through to a time when some better outlet presented itself. He also made a few tie-in deals with VCR manufacturers, allowing them to offer free or discounted cassettes to their customers. And with a sixty-five-thousand-dollar advertisement in *TV Guide,* he launched a direct-mail sales operation called the Video Club of America.

The *TV Guide* investment was a stunning success. From a potential pool of fewer than two hundred thousand VCR owners, nine thousand people joined Blay's club, and the ten-dollar membership fees alone more than covered the cost of the ad. *The French Connection* and *M*A*S*H* turned out to be the most popular titles, but some customers weren't that choosy. "I remember personally taking a phone call in those early days," Blay recalled, "and the guy said, 'I just read about the club and I want to join, but I don't want to wait because I know mail takes a long time. Can you take my Visa number?' It turned out that he was a veterinarian calling from his mobile phone in North Carolina. He gave us his Visa number and joined the club for ten dollars. Cassettes were selling for fifty dollars, and we had only fifty selections. He ordered the entire catalog over telephone. There's a classic case of a guy who had had a dream of having a film library—probably for years."

Spurred on by the *TV Guide* ad, Blay tried other marketing ploys: more ads in national newspapers and magazines; special discounts; buy-two-get-one-free offers. But the results were disappointing. "The lesson we learned was to be patient and wait for a true consumer market to develop," he said. "The initial market was not a consumer market. It was an early-intender market—the people who want to be first on the street with everything."

Fortuitously the Fox titles went on sale just as RCA and Sony were going head to head with their pre-Christmas ad campaigns and just as a new competitiveness—spurred by the arrival of models that bore the Zenith, Sylvania, and Magnavox labels—brought prices under $1,000. As the number of VCRs grew, so did the demand for prerecorded cassettes. By March 1978 Magnetic Video had sold forty thousand cassettes, most of them to retailers who paid a wholesale price of $37.50 each, but a healthy number to individuals who joined the Video Club of America and paid $49.95 each—or $69.95 for a movie that ran more than two hours, since that called for two cassettes. (Blay had decided against using the four-hour speed for his VHS cassettes, since not all VHS machines had it.) By the end of that year he had sold a quarter of a million cassettes, and his production capacity was up to thirty thousand a week, with the factory going twenty-four hours a day.

To avoid becoming too dependent on Fox, Blay looked around for new sources of supply. The other studios were still nervous, but a number of independents were willing to do business. Viacom let him have the *Terrytoon* cartoon series and a group of seven Elvis Presley pictures, and he made an eight-year exclusive deal for much of the Charlie Chaplin library. From Avco Embassy he acquired the rights to ninety-one titles, including *The Graduate*, *Carnal Knowledge*, and *Day of the Dolphin*. There were smaller deals with Lew Grade and Brut Productions, a subsidiary of Fabergé. "So we had an arsenal of films that wouldn't quit," Blay said later. "We had a lock on industry leadership. We could have hired gorillas to sell the stuff."

From the start Fox had intended the deal with Blay to be a temporary measure. The studio retained the freedom to enter the business itself at a later date or to use that option as leverage and try to buy Blay out, making Magnetic Video a Fox subsidiary. As it developed, Blay was willing to be bought out for a price Fox was willing to pay—$7.2 million in cash. The deal was settled in November 1978, only a year after the company had commenced operations. Farmington Hills, Michigan, had its first Hollywood millionaire.

While Andre Blay was staging his fireworks show from a distance, George Atkinson was laying dynamite closer to home. Atkinson was another nobody, but he was a Hollywood nobody. At forty he had already gone nowhere as a movie actor, and he was well on his way toward going nowhere as a small businessman. For more than a decade he had been dabbling in various forms of portable movie exhibition, with unspectacular results, and waiting for the arrival of something like the prerecorded videocassette. "Some people call Andre 'Wilbur' and me 'Orville,' " he said later. "When Andre put 'em out there, I opened my store."

Atkinson was born in Shanghai in 1935. From 1943 to 1945 his family—his English father and Russian mother, a brother, a sister, and himself—lived in a Japanese prison camp. "No big deal—it wasn't Dachau," Atkinson said. "I have a lot of respect for the Japanese."

After several years adrift the family reassembled in Los Angeles in 1949, when Atkinson was twelve. To look at him in his maturity, with his brass-rimmed, tinted sunglasses, his slick sliver-black hair, and his fast-talking, chain-smoking spiel, one would never imagine he had lived anywhere else. Atkinson was graduated from UCLA in 1958, with a degree in English literature and a yearning to act. But after ten years of bit parts and extra work, the fact that Hollywood was spilling over with young men no less good-looking, talented, and ambitious than he sank in. Then he saw something inspiring: a self-contained portable movie-viewing machine that used super-eight film in a cartridge. It had been made by Technicolor as an industrial training and promotional device, but Atkinson saw that it could be used to show old (and uncopyrighted) movies, and he sold the idea as a form of free entertainment to Howard Johnson, Holiday Inns, and Shakey's Pizza, among other clients. When three-quarter-inch video came along, Atkinson graduated to the new medium, hooking up U-matic players to TV sets in bars, and supplying them with cassettes of *The Greatest Fights of the Century* and other such fare.

"I was kind of following the evolution of video," he recalled, "looking for the perfect movie machine, and when I heard about half-inch—about Betamax—I said, 'What's that?' And then *here it comes*—but it's one-hour recording only. I thought, 'Boy, did Akio Morita blow it! I mean, jeez, Akio, baby!' "

After the first two-hour VCRs had arrived and Atkinson read about Twentieth Century-Fox's deal with Magnetic Video, a switch tripped inside his head. By now he had sold his interest in the pizza-movie enterprise, and he was running "a Mickey Mouse little business called Home Theater Systems," supplying super-eight movies for parties out of a six-hundred-square-foot storefront on Wilshire Boulevard, in West Los Angeles. The business wasn't going well. On the personal level, too, Akinson had been through the mill: a divorce and a period of hard drinking which culminated with his living in the back of the store, "saucing it pretty good," as he said later. The Fox announcement brought him to life. If there were people—and Atkinson could testify that there were—who would pay twenty-five dollars for the privilege of lugging a super-eight projector, a screen, and an old movie home for the night, surely there were a lot more people who would pay a few dollars to rent a movie on a videocassette. Atkinson already had his store. To test out his theory, he put an ad in the *Los Angeles Times.* "Video for Rent," it said, though in truth he had nothing to rent. Readers were invited to fill out a coupon and mail it in, "and in less than a week," Atkinson said, "I had about a thousand coupons."

Stimulated by this confirmation of his instincts and subsidized to the tune of ten thousand dollars by an old high school buddy who saw a good investment, he arranged to buy one Beta and one VHS copy of all fifty Fox titles—not from Magnetic Video directly, because Atkinson wasn't big enough to afford the eight-thousand-dollar minimum order, but from a local brown goods dealer who agreed to let him have them for three dollars above wholesale. Then he took out another ad in the *Times.* It was only an inch high and a column wide, but it had the phone "lighting up like Christmas."

Starved for capital, Atkinson established a club. He charged fifty dollars for an annual membership and a hundred dollars for a "life membership." A member was entitled to rent movies for ten dollars a day.

Some of the first customers wanted to buy rather than rent. They were "your Cadillac and Mercedes crowd who would take ten, fifteen, twenty movies at a crack," Atkinson said. Happy as he was to have their money, he didn't have much hope for that side of the business over the long haul. "The studio executives said that Americans are not a renting public—not like the English, who rent television sets," Atkinson said. "They said, 'Americans are an acquisitive people. They want to own.' I said, 'Well, movies may be the exception because of the nature of the beast and the economics of the beast.' I was envisioning a day when there would be thousands of titles, and even if they cost nine ninety-five each and you were Howard Hughes, you couldn't afford to own them all. And I was looking forward to the day when VCR prices came down to match the pocketbook of the blue-collar guy, which would give us a whole new market. A lot of the studio people mistook it. They said, 'Well, people buy records.' I said, 'It's a transient experience. You listen to Beethoven or the Beatles over and over again. You don't watch Burt Reynolds over and over.' "

He was pretty much alone in his thinking. None of the brown goods stores that carried the Twentieth Century-Fox cassettes were renting them. In fact, they had contracts with Magnetic Video stipulating that the cassettes were to be *sold for home use only*. And from the stories Atkinson had read about Fox's decision to enter the videocassette business, he had a suspicion that the studio's executives wouldn't like what he was doing if they knew about it. The possibility that he might be violating the law even crossed his mind, and he decided to make a few precautionary phone calls—to the FBI and the Motion Picture Association of America, among other authorities. Was he or wasn't he allowed to rent? Nobody could say for sure.

In January 1978 Atkinson went to Las Vegas to attend the Consumer Electronics Show, looking to buy more cassettes. Although Magnetic Video was still the only company offering recent Hollywood movies, a few independent operators were selling a variety of odd fare, mostly "adult" titles and old movies whose copyrights had lapsed or fallen into nonstudio hands. In Las Vegas Atkinson made connections with a few of these people. He was also hoping to expand his dealings with Magnetic Video and get out from under the extra expense of buying through a middleman, so he introduced himself to Andre Blay, and "I posed the question to Andre, and he could give me no authority to rent," Atkinson recalled. "He said, 'You'd better talk to Steve Roberts.' And Roberts's response was: 'Very interesting. What would you charge? Would you be willing to let the studio participate?' I said, 'Of course. I mean, I'd like your blessing.' He said, 'Write me a letter.' I thought I was getting the green light."

In the fall of 1978 Allied Artists came out with a line of prerecorded cassettes—a hundred titles, including *Papillon* and *The Man Who Would Be King.* Noting that the Allied Artists contract specifically prohibited rental, Atkinson asked an Allied official if there was "some euphemism we can use, since rental bothers your people." They talked it over, and as Atkinson recalled, "We got cute. 'Well,' he said, 'let's call it a preview agreement. The customer gets to preview the cassette, and if he doesn't like it, he has the right to turn it back to the retailer and gets his money back less a five-dollar restocking charge.' I said okay."

No such authorization came from Fox, but after a few months Atkinson received an unexpected visit from Robert Townsend, the author of *Up the Organization,* who had been retained by Fox as a consultant to study the videocassette market. "He said, 'You've got quite an interesting little operation here, you know?' " Atkinson recalled. "We went and had a Chinese lunch. It was all very chummy. He bought a couple of movies. He might even have rented, I forget. And then I get a phone call a few weeks later: 'Hi, George, how's it going?' I said, 'Great, I'm starting to clone myself. I put a guy into business in Pasadena.' And he said, 'Well, you'd better not rent. Just a friendly tip—you're going to be in a lot of trouble, George.' And boy, I got scared! I had my life savings tied up. Worse than that. I'd gone through a divorce and I was in debt. My whole life was pivoting on my future in video, and here came this dire threat from a big man at a big studio."

The time had come, Atkinson reluctantly concluded, to invest in some legal counsel. "I thought, 'Jeez, I've got a good idea, but if I don't do my homework, I'm going to get my ass chewed off.' " His lawyer researched the question and, by and by, told him, " 'You can't copy it, you can't publicly exhibit it—that's a violation of copyright. But yes, you can rent it, you can eat it, you can destroy it. You bought it. It's your property.' So I didn't have to ask anybody 'cause I wasn't violating anything." Atkinson had alighted on what he calls the "good old doctrine of first sale"—a provision of the copyright law allowing anyone who legitimately acquires a book or other form of copyrighted work to dispose of his particular copy as he wishes. Once the first sale has occurred, in other words, the copyright owner loses control of that copy.

The idea of rental was, of course, in keeping with the Hollywood tradition of selling the *right to see* a movie rather than the movie itself. But with the phenomenal early success of the Fox experiment, studio executives' heads began dancing with visions of vast video libraries in millions of American homes. If people would pay fifty dollars or more for the chance to take Bogart and Bacall, or Redford and Newman, or C3PO and R2D2 down off the shelf at will, what sense did it make to offer them a lower-priced alternative? It made very little sense to Roberts and his colleagues, especially since the first sale doctrine would make it exceedingly difficult to compel retailers to give the studio a piece of any rental action. Rental was an idea to study for the future, Roberts decided, and Blay agreed. For now Fox and Magnetic Video would make their distributors and retailers sign a contract promising not to rent, and anyone who disobeyed would be cut off. This sounded foolproof. But as Atkinson discovered, there was nothing to stop him from "getting my stuff circuitously from another retailer or a friendly distributor who liked selling to me." As a third party he could not be bound by a contract between Fox and somebody else.

Almost from the day Atkinson opened his store, there were customers who looked around with suspicious intensity. Sometimes they couldn't hold it in—the desire to know how they, too, could get into the videocassette business. In the spring of 1978 Atkinson helped one of them start a store in Pasadena. Then, satisfied that his setup was replicable, he decided to franchise it (although, to get past certain regulations governing franchises, Atkinson called the new stores "affiliates"). He placed an ad in the business opportunities column of the *Wall Street Journal,* and because Atkinson himself was busy dealing with his customers, he hired someone to handle the responses. His franchise salesman, Ray Fenster, was given the bathroom for an office. "Ray, I'm sorry, but this is the only space I've got," Atkinson told him. A phone was installed, and "he'd sit there on that goddamn toilet seat and pitch the deal over the phone. Then he'd fly them in to see us. We made it very easy. It was a turnkey operation. We didn't take royalties. They bought the goods from us, and we charged them for licensing the territory. So there was a profit—like a twenty-five percent profit over our costs—but with that you got an education. You know, how do I print up a rental form? What kind of invoice do I use? If you had a certain amount of money—less than ten thousand—we could get you started. And we were selling anywhere from eight to ten stores a month from that toilet seat. Later Ray started up a store of his own, and he became a competitor of mine with big offices in Beverly Hills, and cha, cha, cha! That's what they say about imitation, right? It's the sincerest form of flattery."

Over the next few years the VCR and the prerecorded videocassette set hearts on fire in entrepreneurial breasts all over the land. Americans from every imaginable walk of life cracked open their nest eggs, remortgaged their homes, and put the arm on their parents, siblings, and in-laws in order to become the proud proprietors of Video Castles, Connections, Corners, Hutches, Huts, Palaces, Patches, Places, Shacks, Sheds, Sources, Spots, and Stations. It was as if someone had hung a classified ad in the sky: "Retail Oppty.—Lo Cash / EZ Startup."

Frank Barnako, Jr., a radio and TV producer and announcer in Washington, D.C., had bought a Betamax, and "I saw what a difference a VCR in my home made for me and my wife," he said. "A VCR let us maximize the value of those hours of the day or week that we decided to devote to watching television. I guess after you get above thirty thousand dollars a year or so, you should operate on the assumption that time is your most valuable commodity. If you have a certain hourly value to your employer, you should have at least that hourly value to yourself in your off time, and if you choose to segment some of your off time into recreation, and if that recreation has a component of television watching, it shouldn't just be catch as catch can. And I saw that and I said, "Gee, if I value my own time like that, I'll bet there are others like me.' "

Barnako's thoughts coalesced during a career-planning retreat to which he was sent by his employer, NBC, in 1978. "For two days we were supposed to think about what we wanted to do in life," Barnako recalled. "When I was forced to really think about it, it came down to money. And if it came down to money, I thought I wanted to take a crapshoot on myself. If I can solve problems for these people, I thought, I can probably solve problems for myself. I'd been in broadcasting for fifteen years, and I feared that the next fifteen would be the same, except on a grander scale."

Before Barnako opened his store, he took a trip to Chicago to check out a few of the video stores which were already in operation there. That was a lot more research than many of the early video dealers did before they got started. The typical dealer just found himself a location, cleaned it out, bought some inventory, and waited for the public to beat a path to his door. "I made up my mind to go into business, and four days later I was opened," Jack Messer, a Cincinnati dealer who had dabbled in insurance and real estate, recalled. When Messer took the plunge, in April 1980, the number of available titles was still in the low hundreds. "Twenty or twenty-five thousand dollars bought you every movie there was," he said. It didn't take much space to contain a stock of that size. Messer set up shop in a corner of a friend's record store, with his entire inventory housed in a used display case he bought from a bakery. Magnetic Video and Warner Brothers, which had just come out with twenty-five titles, sent him his initial orders by airfreight. "If you're willing to buy COD, they'll sell you anything," Messer explained.

The rental idea was already gaining ground in Los Angeles and New York, where video stores had existed for a few years by now, but most of the studio executives still expected video to develop into a sales rather than a rental market. So did many of the retailers, including Messer, and his first few months in the business were an education. "In my first month there were six days when I never even got to the cash register because nobody bought anything. People just didn't want to spend sixty to eighty dollars for a movie. All the manufacturers had contracts with the dealers that said you couldn't rent. But as more and more people got into the rental business, it became apparent that they weren't going to try and enforce it. The stores that had rental clubs or trade-in arrangements were doing most of the business, and the stores that refused to rent were going out of business fast."

There were other lessons to learn, and being a pioneer meant learning them the hard way. Before they opened a video store in downtown Phoenix in August 1980, Linda Lauer was a manager for Ramada Inns and her husband, Arthur, built pump engines for deep-water wells. Neither had any previous experience in retail, so the notion that inventory expense ought to bear a certain logical relation to income flow was foreign to them. "Our most serious mistake was not setting a budget for buying tapes," Linda Lauer said. "At first there wasn't much coming out, and then all of a sudden there was a flood of things." By the summer of 1981 Disney, Paramount, Columbia, and MGM (in a short-lived partnership with CBS) had followed Fox and Warner Brothers into the market known as home video. "So we bought everything," Lauer said. "If there was money in the checkbook, we spent it."

With or without retail experience, home video was a hard business to figure. "At any given time about twenty percent of your movies do about eighty percent of your rental business," Jack Messer explained. "So income-wise you might be just as well off if you didn't stock all those other titles or if you sold them off as soon as the demand wasn't there anymore. But in another sense, having a large library sets you apart from the other people in the neighborhood who don't. People say, 'Well, this guy has fifteen hundred titles and this guy has four,' and they're more apt to join the club with more movies. So you can get into a situation where on paper it looks as if you're making a good profit, but if you want to compete, you have to keep pouring that profit back into buying new titles which won't necessarily pay for themselves."

People who couldn't afford to compete discovered that they could be out of the business as quickly as they had got in. "One fellow here spent about three hours setting up his store on Saturday afternoon," Barnako said. "He didn't even nail up shelves to put his boxes on. He decided to sell out two weeks after he opened. Another fellow was in computer software and video software. He called me and said he wanted to get out of the video software. Why? 'Well,' he said, 'I've only got four hundred titles and people go to your store three blocks away and . . .' Damn right! I've got five times that, because that's what the customer wants."

When it became impossible to buy every new movie that came on the market, some dealers relied on their personal preferences or assumed that the top box-office hits would be the top home video hits. Usually they were. "But you'd be amazed," Messer said. "I would have bet that *The Godfather* would be one of the top titles. I couldn't give it away."

In his previous life, William H. ("Doc") Brooks had sold birdseed for a living on behalf of the Hartz Mountain company. But after a few months of running a video store in an Alexandria, Virginia, shopping center, he didn't need any outside experts to tell him what movies were best. "We are a very personalized business," Brooks said. "We've got thirty-five hundred club members and their families. We strive to know them all." Brooks put a suggestion book on the counter and studied his customers' preferences vigilantly. "If somebody walks into my store, I try to guide him to a movie," he explained. "We've got thirty-five hundred people inputting us as to what's a good movie to them, and quite often they're movies that did rotten business when they were in theaters. Or they're movies that got rotten reviews. The movie reviewers don't always reflect what America wants to see."

Strapped for space, some dealers kept their cassettes on shelves behind the counter or in locked display cases, expecting their customers to choose from a printed list. But it turned out that people liked to *look* at the cassettes, or at the boxes they came in, while they made up their minds. And if one store gave them that opportunity, they would patronize it in preference to another that didn't.

Pornography accounted for a large share of the business to begin with, in part because the producers of "adult" movies—unhindered by any prejudice against the new medium—had been its first suppliers. (Well before Andre Blay was selling Twentieth Century-Fox movies for $50 each, *Oui* magazine was advertising Betamax cassettes of Russ Meyer's *Vixen* at $299.95 through the mail, and copies of *Deep Throat,* a movie that had cost $22,000 to make, were selling for as much as $100 each.) As time passed, the allure faded slightly—at least in relative terms—and some dealers found that the presence of X-rated movies in their stores alienated as much business as it attracted. Frank Amato, a Washington, D.C., policeman who opened a video store with his wife and another couple in Olney, Maryland, decided on a policy of strict segregation. "Unless you ask us, we don't tell you we've got 'em," Amato explained. "We've built a separate room—our adult room. No one under eighteen permitted. You go to some video stores in the area and the adult movies are right out there in the open. What's going to happen is, the code of decency is going to come out and say, 'Video stores cannot have dirty movies anymore.' I think it's really bad for stores to have 'em out the way they have 'em."

Home video was a price-sensitive business, but price wasn't everything. "I'm the highest-priced guy in town," Barnako said. "How do I survive? First of all, my stores look great. Everyone working for me wears a tie. Everyone has a college education. I pay them a base plus commission. If you ask my guys a question, you get a civil, intelligent, and often amusing and entertaining answer. We've got colors; we've got purple and gray in our stores. We've got handmade fixtures, custom carpeting. I run a class operation. Our business plan says to our employees, 'We are in show business. We are in that segment of show business that lets a customer watch a movie at home. To get into our theater, they had to pay a heavy admission price: They had to buy a machine.' "

A video store had to keep long hours, and many of them could never have survived without the proprietor's—and his or her family's—willingness to work without pay. "It was two years before I was able to take a Saturday off," Barnako said. "It was two and a half years before I paid myself a salary. It was four and a half years before I was able to give myself a raise."

Doc Brooks kept his store open twelve hours a day and seven days a week for starters, and for a year and a half he was the only paid employee, while his mother and father came in to tend the store in the evenings, after they had finished their regular jobs.

As a police officer Frank Amato had one of the more complicated schedules: "I get up at four-fifteen in the morning so I can report for work at the police department at five-thirty. I get off at two, drive home, and change, and I'm at the store from, say, three o'clock until six-thirty or seven. Then my wife comes in, okay? We have three boys—six, twelve, and sixteen. She brings one of the boys with her—either the twelve- or the sixteen-year-old. I in turn go home to the six-year-old—or she has him with her and I take him home—and he spends a few hours with his dad, takes a shower et cetera, and I put him to bed. Then my wife closes the store at nine o'clock, and she comes home, and we have dinner."

With so much uncompensated labor and so many complicated accounting factors to juggle, it wasn't always easy to determine if a store was making money. "This was a gold rush, this business," Arthur Morowitz, the founder of the Video Shack chain in New York City, observed. "Who comes to a gold rush? People who don't have a lot of estate to leave. People who got fifteen thousand bucks from their aunt and their brother-in-law and every friend they could scrape up. People who had no idea what they were getting into. If they punched a time clock and saw how many hours they were working, they'd probably say, 'This is ridiculous! There's no profit here.' But most of them don't do that. They're having a wonderful time."

Every so often, though, someone would luck into an ideal situation or see something other dealers hadn't seen and turn it into quick profits. As a customer Harry Allen didn't like being asked to pay a membership fee up front. "I resented having to pay a guy to do business with him, and I projected my resentment onto other people," said Allen, who opened up his first video store in a Sacramento, California, shopping center in 1981. "In the rental business everybody thought they had to have a membership—partly as a hedge against theft but also because a lot of people were undercapitalized and they needed cash. When we opened our store, we decided we wouldn't have a membership, and I think that accounted for a large part of our success."

Allen had been teaching English for a living. He saw the store as a part-time thing which his wife could oversee. "It started out as her venture—it was going to be our second income," he said. "We opened with only about a hundred tapes, which was ludicrous. But we had a good location, and the other stores in the area—there were three or four within a mile radius—all had memberships. In two weeks we were in the black. Within about six months the store was making a lot more than I was as a college teacher, and I could see the writing on the wall." He took an indefinite leave from his job at Sierra Community College. "I'm still trying to decide what I want to be when I grow up," Allen said four years later. "But I probably won't go back to teaching. It's heresy for someone with an English background to say, but I don't think that writing right now is the major way for people to inform themselves. I enjoyed teaching, but . . ." His thought trailed off into the unspeakable zone of cost-benefit analysis. "You know," he added, "they could always put my English class on videotape. I used to pretend that it was a discussion class and not a lecture class, but I was lying. It was a lecture class."

THE RENTAL WARS

IN THE SPRING and summer of 1980 Morton Fink had reason to feel pleased with himself. A former vice-president of Sonam, Fink was now the president of Warner Home Video, a company established by the entertainment conglomerate Warner Communications to market movies on videocassettes. He regarded himself as a "start-up specialist," and he was proud of what he had accomplished in his relatively short time with Warner. It had taken him only six months, from when Warner had given him the go-ahead, to bring his first twenty-five titles to market. They included *Superman, The Exorcist,* and *The Searchers,* a personal favorite. Being a movie buff and a believer in the importance of small touches, Fink had given the Warner cassettes a distinctive appearance. They opened up like a book and came with cast lists, production credits, stills, and liner notes printed on the front and back covers. Fink intended to make video shopping a "tactile experience," and his efforts had been noticed. "We were the hottest booth exhibiting at the Consumer Electronics Show in Las Vegas that January," he recalled. "We did an enormous amount of business."

Already Warner Home Video was more than an outlet for Warner Brothers movies. Fink had licensed home video rights from New World Cinema and Filmways, and he was working on a sixty-million-dollar deal with United Artists. His ability to negotiate on that scale was a sign of the company's bright place in the Warner galaxy. So was the fact that Fink reported to the office of the president, directly under Steven Ross, the chairman of Warner Communications. "We were using the movie company's assets," Fink explained, "but if we were part of the movie company, there was the danger that we would do things solely in the studio's interest. And if we reported to the record people—although we used them as distributors—we would be influenced strictly by what was in their interest. Steve Ross recognized that this was a business that would take elements of publishing, records, and movies, and he wanted it to be independent. So he set it up that way, and it was understood that I would meet on a regular basis with Warner Brothers to talk about our release schedule and how the business was going."

But when Fink walked into a meeting with the studio people one early summer day in 1980, he found them in a restless mood, and with more on their minds than which titles would go out on videocassettes next. The way Frank Wells and Ted Ashley, the studio's president and chairman respectively, looked at it, Warner Home Video was an interesting experiment which had been allowed to proceed on its merry way for a year, and Fink was not a bad fellow for someone with offices in New York and a nonmovie background—for a suspicious character, in other words. But the time had come to get serious. As Fink recalled the meeting, "They said, 'We've given you X number of pictures, and you've paid us nice royalties. But . . .' " It was a big "but." Like Fox, Warner had expected people to buy its movies rather than rent them, and it had expected most of the sales to be through appliance stores, department stores, or record stores—businesses where videocassettes would be a sideline. Instead, the studios had found themselves dealing with the "video software dealer," a species of small businessperson whose existence they had never contemplated, and the market had gradually shifted away from sales toward rentals. The typical Warner cassette, at $79.95 retail, brought Warner Brothers a royalty of something like $10—a handsome return if multiplied by millions of individual purchasers, but a far less attractive one when multiplied by ten or fifteen thousand video store owners, each of them free to take a cassette which had cost, say, $50 wholesale, and rent it out a hundred times or more at $5 a shot.

To Wells and Ashley and their colleagues, it seemed as if *their* share—the creative share—was pittance compared to the money that was flowing to the retailers, the VCR manufacturers, and just about everybody else involved. "The studio," Fink explained, "was saying, 'Look, we're not getting any part of the rental, and the retailers are buying two VHS and a Beta of each picture—and they're probably duplicating a couple more in the back room, which is piracy from a legal point of view, but you have to find it.' "

The public, too, seemed to be getting a big return on a small investment. Already a number of stores had lowered rental prices to two dollars. Even one-dollar rentals were not unheard of. As the number of VCRs grew, home video was bound to become a more important element of the movie business—conceivably at the expense of the traditional movie theater—and the thought of masses of Americans paying a dollar or two (per family, or viewing unit) to see movies that cost ten or twenty million dollars to produce . . . Well, it was more than any red-blooded studio boss could bear. The phrase *crown jewels* was mentioned. Warner Brothers, Fink was told, was "giving away its crown jewels."

"Why can't you license motion pictures to retailers just the way we license them to theaters?" Wells asked Fink. It was more than a friendly suggestion. The message, as Fink received it, was: "Unless you come up with a business system that allows us to participate in all the rentals that take place, you won't get any more product."

If a true pioneer of the movie business had been in the room—someone whose memory went back to the Edison and Biograph days—he might have told Ashley and Wells, just as a point of information, that history was repeating itself. He might have explained that the movie companies had made a similar discovery in 1908: Wholesalers and retailers were getting too big for their breeches by buying movies at flat rates and selling viewing rights to a rapidly growing audience. And they were enhancing their earnings by engaging in insidious practices such as mixing legitimate (that is, Trust-approved) movies with bootleg fare. Edison and the other leading producers, the pioneer might have recalled, had said many of the same things then that Wells and Ashley were saying now. They, too, had decided to end the practice of selling movies and to construct an elaborate licensing system designed to impose order and bring a rightful share of income back to them. After years of struggle, in which most of the original companies had gone under as a result of various economic and artistic miscalculations, the industry had more or less managed to fulfill that goal. And now, the pioneer might have said in conclusion, it seemed that a handful of young Hollywood executives had given away that hard-won achievement in the space of a couple of years, for a quick buck.

But there was nobody of that sort in the room. There was only Morton Fink, and he had two reactions to Wells's and Ashley's position. One, it sounded reasonable. Two, he didn't want to be pressured into something he might regret. So he did what any self-respecting corporate executive would have done in the circumstances: He initiated a study. To conduct it, he retained Leon Knize, a marketing and distribution consultant of his acquaintance. The first priority, Fink and Knize agreed, was to get a clear picture of the home video business as it stood. They had to have reliable statistics on such questions as what it cost to run a video store, the size of the demand for popular titles and unpopular ones, the ratio of sales to rental transactions, and the amount of inventory a store reasonably needed to possess.

Knize spent six months gathering information, largely by interrogating the retailers who carried Warner Brothers movies. His findings confirmed the studio executives' suspicion that the great majority of transactions at the retail level—well over ninety percent, it turned out—were rentals rather than sales. Knize also determined that a high percentage of the rental business was done by a relatively small number of movies—basically the six to twenty top hits that Hollywood produced in a year. Yet even with those titles the demand tailed off after the first two or three months. So a lot of income hinged on the retailers' ability to choose the right titles in the right quantities at the right time and on their having deep enough pockets to afford them. (Weston Nishimura, the founder of a Seattle video store chain, explained this dilemma from a retailer's perspective: "With some movies you could order twenty copies and it's not enough on the first day. Three weeks later it's one too many. A month later it's ten too many. What do you do?") Short of capital and scared of making a mistake, most dealers bought as many titles as possible, but only a few copies—say, two VHS and one Beta—of each. And the result, in Knize's judgment, was that no one—neither the dealers nor the studios—was realizing anything like the full potential of the business. In fact, most dealers weren't making any money at all.

"I came to the conclusion, correctly or incorrectly," Knize said, "that the business at the retail level wasn't profitable—unless the percentage of income that came from what we might call 'mysterious merchandise' or porno, or 'mysterious merchandise slash porno,' was making up the difference. The honest dealers were losing money, unless they had tremendous volume and the broadness of their libraries was more effect than reality, or they had multiple stores and could take a narrow inventory and spread its effects."

So the Warner Brothers executives were wrong, in Knize's view, to think the retailers were getting the better of them. The retailers had problems of their own, although in their enthusiasm they might not be willing to face up to them. But the studio people were right in wanting an arrangement in which videocassettes would be leased rather than sold, and that idea, Knize concluded, made sense for the retailers as well as the studios.

From Warner's point of view, a rental plan would make it possible to release videocassettes region by region, just as the studios did with theatrical movies. This would allow Warner to get more copies of each title into stores at the time of the peak demand, while getting by with fewer copies overall. (A computer model was developed to predict the demand for various categories of movies and to map out an appropriate release plan for each.) As for the retailers, they would be able to order as many copies of a new title as they wanted, without making a big financial commitment up front and without worrying about what would happen to the value of their cassettes three or four months down the road. Whenever they thought they had even one cassette too many, they could simply send it back to the studio.

With more copies of the top titles to rent out, the retailers would experience a sharp increase in the number of rentals, Knize predicted, and everyone would benefit. But he acknowledged that it would be hard for one company to set a new course while the rest of the industry stuck to the old one. For the Warner approach to succeed, thousands of retailers had to be persuaded to stock more copies of new movies. What if they persisted in their established pattern of taking a few copies of every title? The fact that they weren't doing very well financially under the present system, in the aggregate, did not mean they would leap at the chance to try something different. As Fink explained, "They could always do without one studio if they wanted to."

In 1980 Disney had tried a two-track marketing system, with some cassettes designated rental only (and packaged in blue) and others designated sale only (and packaged in white). Disney had sought to enforce this distinction through contracts with a network of "authorized" retailers. But many retailers had gone right on renting the sale cassettes, and though Disney had never formally abandoned the plan, it had found itself powerless to enforce it. "The retailers had already become used to living in a world of anarchy where they could do anything," James Jimirro, the president of Walt Disney Telecommunications, observed. "If all the studios, from the beginning, had made cassettes available for rental only, the retailers would have cheered. I remember saying to George Atkinson, 'George, if Fox had come to you and said, "We're going to license these things to you for rental," what would your response have been?' And he said, 'My response would have been, "Fine, great." ' And yet in no time he was saying, 'The only way this business can work is the way it's worked over the last eight months.' "

Warner was hoping to do nothing less than lead the entire industry away from a market structure posited on sales toward one posited on rentals. To get that move started, Fink and Knize believed that they would need the backing of at least one, and maybe two, of the other studios. This was a tricky requirement. Unlike Thomas Edison and his cohorts, who had claimed exclusive patent rights over the equipment used to make and show movies, Warner had no legal basis for telling the other movie companies how to behave. Even if the Warner people had thought they could win over their fellow studio bosses by power of persuasion alone, they had to consider the antitrust laws, which (in the United States, unlike Japan) barred an industry's leaders from formulating their policies in concert. On the other hand, Fox was known to be dissatisfied with the way home video was going, and it, too, had hired a consultant to study the problem. Warner and Fox had something else in common: They both had home video deals with United Artists, Warner for foreign rights, and Fox for domestic. When Fink pointed this out to Steve Roberts over dinner one night in Beverly Hills, Roberts appeared surprised. "Congratulations," he said. "I thought I had the foreign, too." From there they got to talking about the advantages of size in the home video business and, before long, about the idea of a merger between Fox, with its United Artists package, and Warner, with *its* United Artists package. "And we shook hands that night, pending the approval of our mutual managements," Fink recalled. "I then thought to myself, 'We're going to have that sixty percent of the important titles we need to make rental work.' And on the strength of my belief that we were going to merge these two companies, I launched my rental program in the state of Texas as a test."

In retrospect, it occurred to Fink that he might have postponed his test until the merger actually happened. But on September 2 1981, Fink called a press conference to unveil the Warner Home Video Rental Plan, which he described as "a carefully programmed response to consumer demands and . . . a natural action of the marketplace." The home video business had experienced "an irreversible, dramatic, consistent trend toward rentals," Fink explained to an audience of entertainment reporters. The existing structure of the business, which compelled retailers to pay a relatively high price for each cassette and be stuck with it forever, made it impossible to respond to sudden fluctuations in consumer demand, he said, and imposed an unacceptable burden of financial risk on people who could not afford it. Henceforth, Fink said, Warner would assume the risk. Dealers would be allowed to lease cassettes on a weekly basis—at rates that declined over a six-week period from $8.25 to $4.40—and return them whenever they chose. It was an arrangement calculated to make the studio's earnings on each title reflect, in a rough way, the rate at which the stores succeeded in renting out that title. (Actually to collect a payment for each rental would be prohibitively complicated, Warner had concluded.) The plan would benefit all parties, and there would be no turning back, Fink declared. "We will no longer sell our product to anyone," he said.

The atmosphere at the Warner offices in New York could not have been balmier. In the great state of Texas, however, there was "skepticism aplenty," according to an account in *Video Week,* and as further details of the Warner plan surfaced, stronger emotions took over. It seemed that the plan entailed a recall—to be followed by a reissue in rental-only packaging—of certain titles currently available for sale. This struck a lot of dealers as dirty pool. In the Dallas area thirty or forty of them held a meeting to discuss the plan and voted overwhelmingly to reject it unless Warner made major amendments. The take-it-or-leave-it basis on which it had been presented "created a lot of hostility," one retailer said, adding, "The city of Dallas is against it." The recall provision was "totally unacceptable."

George Atkinson, speaking on behalf of three hundred Video Station affiliates, including a few in Texas, called for a de facto boycott of the "Warner Rocky Horror Rental Show." His stores could get by without Warner titles, Atkinson said, so "it's not the end of the world."

Abruptly Warner shifted into apology mode. In Fink's mind, the problems were minor and involved details like the recall—which was promptly modified—rather than the basic concept. "It was a very well-conceived program," he said later. "But it was dramatic—and it was different." Chastened by the Texas reaction, Warner set out to calm the retailers' fears and get them at least to try the plan. A public relations firm was hired to prepare a white paper explaining the benefits.

While Fink was trying to smooth things over with the retailers, his courtship of Fox was running into trouble. The dinner with Steve Roberts had been followed by a meeting of their respective bosses, Steve Ross and Alan Hirschfield, and the urge to merge had come upon them as well. "Let's make it happen by Thanksgiving," Hirschfield had said. Then the question had climbed another rung up the Fox ladder: to the studio's principal stockholder, the Denver oilman Marvin Davis. It so happened that Davis, around the same time, was trying to free up some of Fox's studio space for a real estate development deal—a project that threatened to leave Fox with a shortage of sound stages. Davis mentioned that problem to William S. Paley, the chairman of CBS, who had some underutilized studio space in the San Fernando Valley. Paley mentioned a little CBS problem: the network's desire (piqued by a brief alliance with MGM) to get into the home video business in a serious way, despite a lack of videocassette-manufacturing facilities and distribution machinery. Davis and Paley decided that they could solve each other's problems by joining CBS's studio space and Fox's home video company into a new entity, to be known as CBS/Fox Video. And Warner, suddenly, was out of the picture.

So in his dealings with the retailers, Fink found himself obliged to use more carrot and less stick than he had intended. But Warner would not stand entirely alone. In November Fox announced a rental plan of its own devising. Less drastic than Warner's, it created a six-month "window" when new titles would be available to retailers for lease only—at seventy-five dollars for hits and forty-five dollars for everything else. The retailers, in turn, would be free to rent the cassettes to the public on whatever terms they chose. Fox went out of its way not to send them into shock, as Warner had. The six-month period made Fox's plan much less complicated than Warner's with its week-by-week leases. And though Fox reserved the right to recall a title when the six months were up, it held out hope that retailers would be allowed to pay a small additional fee for the right to hold on to a cassette and either continue to rent it out or sell it off as used goods. The Fox plan won endorsements from a number of retail chains. A few California retailers announced a boycott, but George Atkinson, notably, was not among them. He had analyzed the Fox plan and concluded that it was "not dangerous."

Fink announced that he was "thrilled and delighted" by Fox's plan—even though it differed from Warner's—because "another studio is standing up and being heard."

In December MGM / CBS—an alliance that would soon dissolve so that CBS could link up with Fox—launched a rental plan that was similar to Fox's but built around a four-month instead of a six-month term. A trend seemed to be under way—a trend that Warner had set in motion. Yet Fox and MGM refused to go as far as Warner had gone; they continued to make most of their movies available for sale to retailers, and even the top hits, they indicated, would go on sale after a period of months in the rental-only category. Warner, unlike Fox and MGM, was facing large-scale boycotts, and it began to dawn on Fink and his colleagues that they might have overreached.

"The retailers wanted to own," Leon Knize explained. "That was something we didn't fully anticipate."

A cassette that was owned was, for one thing, a cassette that could be sold, and while sales to individuals—the sell-through market, as it was called—might account for only one in twelve transactions, the profit from one sale could be worth almost as much to a retailer, in dollar terms, as twelve rentals. Ownership had other, more indirect advantages. It gave the retailer an inventory that could be depreciated for tax purposes or borrowed against in time of financial need. A lease arrangement, on the other hand, left the studio with those privileges. "Right now if you buy a thousand tapes for forty thousand dollars, you can argue one way or another with an accountant, but you have forty thousand dollars' worth of product on your shelf," Frank Barnako explained. "You have something that you may be able to get a bank to give you twenty percent on if you need some more money. If all the stuff that's on your shelf isn't really yours—well, what's better, renting a house or buying a house? That's the difference from a financial standpoint."

The new year brought a new Warner rental plan, aimed at bringing Warner more in step with the other studios, without any change in its "strategic purpose." The new plan, called Dealer's Choice, provided for a rental term of either four weeks or six months, whichever the retailer preferred. Like Fox, Warner would have "A" and "B" titles; the former would cost eighty-four dollars to lease for six months and twenty-two dollars for four weeks, and the latter forty-two dollars and eleven dollars. Warner would also create a third category—with more than eighty percent of its titles—of films that would again be available for sale (or "lease / purchase," to use the Warner euphemism) on easy monthly credit terms. Warner calculated that retailers would owe less money under its system than under either the Fox or the MGM arrangement.

"We're dedicated to making this business work from the long-term point of view," Fink said at the press conference he called to unveil the new plan. But to get to the long term, one must survive the short term. Warner had timed the announcement of Dealer's Choice to coincide with the opening of the Consumer Electronics Show, a major gathering place for the nation's video dealers (many of whom dealt in VCRs as well as videocassettes). Confident that they had ironed out the kinks in their plan, Fink and Knize came to Las Vegas in a hopeful—but anxious—mood.

Similar emotions were at work in the video dealers. They had been through hard times, when the scarcity of product and the scarcity of potential customers—VCR owners—had made home video a marginal business. Now the outlook was improving. Some of the pioneers were starting to pay themselves salaries; to take a few days off every week; to work at a desk, executive-style, instead of chained to a cash register; and maybe even to think about expanding. VCRs were selling at a fast clip; in 1981 alone, Americans had bought over a million of them. Noting the early resemblance of the VCR sales curve to the color TV sales curve of twenty years past, people in the industry were predicting that in five or ten more years the VCR would be part of the standard equipment of middle-class American life. If a retailer could come even close to making a profit in the present circumstances—with VCRs in only three percent of the nation's households—how could the business be anything but a bonanza when household penetration reached, say, twenty or thirty percent?

But just as this charming prospect beckoned, the studios had begun to rewrite the rules. No matter how much they insisted they were only trying to improve the business for everybody, the fact remained that the rental plans—all of them—made it impossible for a retailer to buy a hit title at a flat price when it first appeared. However the studios explained the rationale behind their plans, it looked to the dealers as if the underlying purpose were to capture a bigger share of the take from each rental. They felt like colonists who had been through a few tough winters and were about to harvest a bumper crop only to learn that the mother country was going to raise their taxes.

To most of the video dealers the movie studios were remote institutions run by people who didn't understand how the business worked at the retail level. As Frank Barnako explained, "*Superman II* from

Warner's was twenty-two dollars for four weeks. Lose a day at either end for shipping, and you've got twenty-six days to make it back in. Now let's say that at that time we were getting five dollars for a three-day rental. So twenty-two dollars comes back to you in approximately five turns. That's fifteen days, and it leaves me with eleven days to make a profit in. Doing as much business as I possibly can, I could hope to fit in four more rental periods of three days each—that's another twenty dollars. So everything else being equal, I'm going to make twenty dollars on twenty-two. That's less than a hundred percent. But everything else is not equal. I have defects. I have customers who say, 'Yes, I'll pick it up,' and they don't, and I lose that night. Never mind shoplifting; never mind defective cassettes or customers whose machines eat the cassette in the first day of the rental period. Never mind the customer who switches the cassette and gives me back *Roots* instead of *Superman II.* Those are all very real retail problems, but around the board tables in New York, where the decisions about video are made, nobody cares about them. If a customer is a day late or an hour late and has some excuse, I guess I'm supposed to say, 'Tough luck. You owe me another five dollars.' Well, if *I'm* late, the distributor can treat me like that because I've got nowhere else to go. But if I get too rough with a customer, the guy has got thousands of other places to go. So the customer is always right. Warner Brothers didn't seem to understand that. To them it was just a business. It was our lives."

The Warner plan stood out from the other rental plans because it was the most complicated. "I'm sure it looked good on a computer screen," Barnako said. "It might even have been manageable in an office staffed by people who were educated past high school, whose honesty was unquestioned. But here you were dealing with small businesspeople—moms and pops—and if they had any employees, it was probably a kid making minimum wage. And we didn't just have one rental program. Counting Disney, Fox, and MGM, we had four. Each one had different expiration dates. Each one had a different number of forms."

Rocco LaCapria, the proprietor of a video store in the Bay Ridge section of Brooklyn, didn't know much about Hollywood, but what he knew he didn't like. "If you look back and see what Hollywood has done to people like Mickey Rooney and the mom-and-pop theaters, you know they're ruthless," LaCapria said. "All they want to do is make money." LaCapria had been in the video business for barely a year, but he had already built an association of dealers—125 of them—in the New York / New Jersey / Connecticut area. Now, along with dealers from Chicago and Southern California, he was organizing a national association. When it met for the first time, in Las Vegas, anti-Hollywood resentment was at the top of the agenda.

"There's no leasing program we want to accept," LaCapria declared to cheers from about two hundred dealers.

"These are the same studios that told us four years ago rentals wouldn't work," Cheryl Benton, the executive director of the Video Station stores, told the group. "And now that they see that it does they're saying, 'Oh, good, let's get our fingers in that one, too.' "

"Cable is getting movies before we do," another dealer protested. "Warners is plastering them all over cable. They're trying to sponge us. They think of us as the shortchange department."

Bob Brown of Video City in Oakland, California, wore a T-shirt that said, "Boycott Warner Home Video, Superman, Private Benjamin, The Shining."

But a boycott didn't seem like much of a protest to some of his comrades. "Let's pull their booth apart," one dealer suggested.

"Rip their catalogs apart, and throw them in their faces," another voice was heard to cry before LaCapria reclaimed the microphone and called for order.

Whatever else the Warner plan might accomplish, it had helped create a sense of solidarity among the dealers. Not just one but two national organizations were being formed in Las Vegas that winter: LaCapria's Video Software Retailers Association and the rival Video Software Dealers Association. The policies of the VSDA, led by some of the larger retailers like Nishimura and Barnako, were more moderate than those of the VSRA. Still, the Warner plan was something both groups could agree on; they detested it.

The Las Vegas experience stimulated Leon Knize to undertake a campaign of shuttle diplomacy, meeting with hundreds of video dealers around the country. If they understood what Warner was up to, he believed, they wouldn't be so suspicious. "We were trying to do what was best for the dealers," he said later. "It was an idealistic plan, and we did not wish to—nor could we—use anything but persuasion to sell it."

Having sized up Rocco LaCapria as a man of influence in the retail community and a man who did not yet seem to appreciate the innocence of Warner's intentions, Knize decided to pay him a visit. As Eisenhower had gone to Korea, Knize would go to Brooklyn. One winter day in early 1982 he spent three hours hashing things out with LaCapria and a fellow dealer. Later Knize took them and their wives out to dinner and pressed his case further. Still later he brought a Warner associate with him to a meeting of LaCapria's organization, attended by roughly seventy-five dealers from New York and New Jersey. The meeting was held at the Yacht Club in Bensonhurst. (Gazing inside at the spectacle of "a bunch of guys with leather jackets," Knize's associate got cold feet. "Leon," he said, "are we really going in there?") Knize came armed with an awesome store of statistics, and he coolly laid out the evidence in black and white for the proposition that most video dealers weren't making a profit. He explained about the "demand pyramid" and the problem of "inventory risk." He assured his listeners that a lot of what they had heard about the Warner plan was untrue—or, at any rate, no longer true—and he spelled out the reasons why the plan was bound to benefit them as much as the studio.

"Some of the younger dealers did not give the man the respect that he was due," LaCapria said later. Others listened politely and were inclined to give Knize high marks for valor under fire. But very few of them could have reconstructed his presentation on an argument-by-argument basis. The Warner plan was sophisticated. The retailers, by and large, weren't. "Your graphs are great," LaCapria told Knize. "They can prove whatever you want them to prove."

At about the same time, in Phoenix, Warner's emissaries were pleading with Linda Lauer to try the plan. "They begged us to take it because we had three or four stores by then," she recalled. "They even asked us to take some cassettes on a trial basis, with no charge, to show us that the plan would make money for us. We refused. I said, 'Nope, I don't want any part of it.' We had very heated arguments. They just could not understand why we were not partaking of this. We could make money doing this, they kept saying. We should try it. And we said no, thanks. We never signed up for any of the rental programs—not a one—'cause we didn't believe in 'em. Most of the rental plans said, 'You can rent this movie for six months, and at the end of six months you can buy it.' So we just waited for the six months to expire and bought it. And we told our customers exactly what we were doing and why. We felt that the manufacturers were trying to regulate this industry by telling the consumer what they could and could not have, and we didn't feel that was right. And our customers felt the same way. Now certainly there were stores in our area that did sign up for the rental programs and had tapes sooner than we did. We told our customers, 'You're more than welcome to go down and rent those tapes if you don't want to wait for us to get them.' It didn't seem to make that much difference. It didn't hurt our business."

One thing Linda Lauer positively could not abide was the right asserted by Warner in its contracts to make unannounced inspection visits to participating stores and to sift through their records and inventory. "If we signed an agreement, they could come in and audit at any time," she said. "It gave them access to my books. The only people I feel that have the right to audit my books are the IRS." Even without that provision, though, she would probably have rejected the plans because "I did not want to have product in my store that I did not own."

That feeling proved to be more widespread, and intense, than the studios had bargained for. Maybe it was irrational to buy cassettes outright and be stuck with them permanently when the demand was so transitory. Maybe it was irrational to pay the same price, generally, for every picture, from the hottest to the most shopworn. But it added a gambling element to the retaliers' life. You could win big or lose big with each decision, and many video dealers liked that. The inventory risk which the studios wanted to lift from their shoulders was a burden they were perfectly happy to bear. In fact, they insisted on it.

"The reason people go into small business is to have some control," Weston Nishimura explained. "If you look at the history of Hollywood, it's been a history of trying to have disproportionate control—complete vertical control."

Despite all the diplomacy that Warner could muster, hundreds of video stores went right on boycotting. Others signed Warner's contracts but simply didn't order very many cassettes. They had discovered that they could make money by leasing the top few titles for a few weeks—"when they were most current, most hot, most promotable," as Arthur Morowitz explained. "But then," he added, "Warner couldn't make any money." Gradually Warner put fewer titles out for rental and more out for sale. A parallel pattern could be observed at Fox, and by the summer of 1982 both studios' rental plans were history. "We couldn't fight the tide," Fink explained.

MGM fought on for a few months longer, although most of the dealers who took *Tarzan,* the first title in the MGM rental plan, failed to return it when the rental period of 120 days was over. William Gallagher, the vice-president in charge of MGM's home video operation, said he was "willing to chalk up" the dealers' negligence "to some trial and error." But he could find no excuse for their behavior with the next title, *S.O.B.,* so he wrote them a stern letter. "Your adherence is so poor that it can only be characterized as ridiculous," Gallagher informed the delinquents. If they didn't shape up, he added, they risked being cut off. "Your pride is on the line," he warned.

The MGM rental plan was also on the line. Only three weeks later MGM dropped it. The dealers hadn't "reached the maturity" necessary for such a plan, Gallagher said.

Immature they might be. Ineffective they weren't. Their ability to mount a coordinated campaign of resistance from coast to coast was a source of amazement and annoyance to the studio people. "A thousand small retailers can band together in an 'active boycott'—that's their word, not ours—with impunity," said Lawrence Hilford, who assumed the post of president of CBS / Fox Video in 1983. "But if two studios are seen on the same side of the Mississippi on the same day, they will get accused of restraint of trade."

HOLLYWOOD ON THE POTOMAC

ON OCTOBER 21, 1981, the day after the Ninth Circuit issued its decision against Sony, Stanford Parris, a Republican congressman from northern Virginia, denounced it as "the latest example of idiocy in the federal judiciary" and introduced a bill designed, he said, to "halt an unnecessary and unwarranted intrusion by the federal courts into the private homes of three million Americans." Parris was one of those three million. He had bought a VCR for time-shifting purposes, with special attention to Washington Redskins' games. His bill was one sentence long and took about an hour to draft. It said that "the fair use of a copyrighted work includes the recording and use of copyrighted works in a private home by an individual on a home videorecorder if the recordings are not used for direct or indirect commercial advantage." It was not, perhaps, the most thoroughly researched or carefully crafted piece of legislation ever proposed. But there are times when the conscience of an elected official tells him that he must move boldly and quickly—or some other elected official will get in there ahead of him. Indeed, several members of Congress had seen the urgency of the situation. In the House John Duncan, a Tennessee Republican, offered a similar bill the same day, as, in the Senate, did Dennis DeConcini, an Arizona Democrat, with the cosponsorship of

Alfonse D'Amato, a New York Republican.

Out in the wider world, the decision met an equally swift and emphatic reaction. "It's so stupid it hardly bears comment," declared an editorial in the *Kinston* (North Carolina) *Free Press.* "Big Brotherism," said the *Salt Lake City Deseret News.* "Something misfired in the judicial process," concluded the *Wall Street Journal.* Enforcement would make "Prohibition look easy," opined a columnist for the *Los Angeles Times.* The *Martinez* (California) *News-Gazette* contended that the decision, "carried to its logical extreme . . . would forbid singing a copyrighted song to one's self in the shower." A cartoon in the *Philadelphia Inquirer* showed a VCR and a revolver side by side and asked the question, "On which item have the courts ruled that manufacturers and retailers be held responsible for having supplied the equipment?" Countless cartoonists and columnists—and Johnny Carson, in a sketch for "The Tonight Show"—imagined frantic households being visited by "Video Police" or detectives from the "Betamax Squad." Tom Shales, the TV critic for the *Washington Post,* imagined himself as an underground videotaper in the dread year 1984. "It is the third year of our persecution, oh my brothers, oh Betanauts and Betalogues and fellow members of the Betanese Liberation Front," he wrote. "Still, they come, the gray men in the greatcoats, the storm troopers in their clodhoppers. . . ." In the long history of its offenses that much-maligned branch of government the judiciary had probably never issued a decision that attracted more abuse and less sympathy.

The cartoons, the editorials, and, above all, the bills led to a sharp increase in the volume of phone traffic to and from the office of Jack Valenti in the headquarters of the Motion Picture Association of America, just a few blocks up Sixteenth Street from the White House. As a onetime inhabitant of that other presidential address and a graduate of the Lyndon Johnson school of meteorology, Valenti knew a typhoon when he saw one. Although the Ninth Circuit's decision could not take effect for a minimum of several months—first the Supreme Court would have to decide whether to hear Sony's appeal—DeConcini and Parris and their allies were treating the issue as a crisis of state. "We had to counterattack immediately," Valenti said later, "or else our court decision would have been nullified. We would have been overrun congressionally by the DeConcini and Duncan-Parris bills." There was a frightening precedent. Once before Stan Parris had become worried about a threat to the people's right to watch the Redskins, and he had proposed a bill outlawing the National Football League's TV blackout policy. His bill had made it through the legislative process in near-record time: seventeen days from introduction to passage.

What Valenti needed, and needed fast, was a palatable alternative to the home recording exemption—some formula that would allow Congress to satisfy the clamor for action without completely undoing the Ninth Circuit decision. The decision itself hinted at the answer. The court, while accepting the plaintiffs' reading of the law, was clearly skeptical about the wisdom of ordering the Betamax off the market. Citing *Nimmer on Copyright,* a standard text by Melville Nimmer of the UCLA Law School, the Ninth Circuit had observed that there are times when an injunction may do "great public injury" and when a royalty arrangement "may very well be an acceptable resolution." (In view of these comments, the Video Police state depicted by the cartoonists and editorialists was, perhaps, an alarmist creation. "It's too bad—I wanted to see the uniform," one copyright scholar, Peter Jaszi of American University, remarked at the time.) A royalty on VCRs and blank tapes, it seemed to Valenti, could be promoted as a compromise which would preserve the interests of consumers and creators simultaneously.

Sidney Sheinberg had disliked the idea of a royalty when Harvey Schein proposed it five years earlier, and he still did. He especially disliked the idea of asking Congress to do the job, as Valenti was urging. Congress, he feared, would give the copyright owners' interests short shrift rather than pass a bill that could be portrayed as "hurting the consumer." The problems of cable TV and jukeboxes had been resolved by congressionally imposed royalties, and in both cases the sums of money involved struck the people on the receiving end as preposterously small. (The annual fee for the privilege of maintaining a jukebox on one's premises was set, initially, at eight dollars.) The royalties paid by nightclubs, TV variety shows, and other incidental users of copyrighted music to the musical collection societies ASCAP and BMI were a more pleasing precedent, but they had been fashioned, without congressional help, in negotiations among parties that agreed on the basic legitimacy of the copyright owners' claims. The judicial process, it seemed to Sheinberg, offered more hope than an appeal to Congress.

On the other hand, he could no longer seriously hope to send the Betamax back where it came from. He and his company had come to understand the VCR's appeal in the most painful way—by watching their own entry into the home entertainment marketplace, the laserdisc, get trounced. In theory, the laserdisc offered a superior picture, but the picture was not superior enough to impress most people as sufficient compensation for the inability to record. In theory, too, the player and the disc were less expensive to make than a VCR or a videocassette. But with a four-year head start the VCR manufacturers had achieved formidable economies of scale, and the retail price of the average VCR had dropped from fourteen hundred to seven hundred dollars—about the same as the price of a laserdisc player. Ads for laserdisc players lured a goodly number of customers into appliance stores, but many of those customers walked out with VCRs instead. By the end of 1981 MCA's laserdisc subsidiary, DiscoVision, had laid off most of its employees, and the parent company was preparing to bail out of the business, with cumulative losses in the neighborhood of thirty million dollars.

Shortly before the Ninth Circuit decision Sheinberg had given the go-ahead for MCA to enter the prerecorded videocassette business. He had avoided that move as long as he could, both as a matter of principle and to avoid muddying the legal waters. If Universal had been selling its movies on videocassettes at the time of the hearing before the Ninth Circuit, Dean Dunlavey would undoubtedly have made much of the fact. Now, however, the home video business had grown too big to ignore, and as long as Sheinberg was resigned—and commercially committed—to the VCR's existence, he had to admit the strength of Valenti's argument that a hard-line stand by the movie industry would add to the allure of the bills introduced by Parris, Duncan, and DeConcini, which threatened to make the lawsuit moot and to leave Hollywood without any compensation for home recording.

In Twentieth Century-Fox, Valenti faced another challenge to his diplomatic skills. As the first of the studios to market its films on prerecorded videocassettes Fox was bullish on home video. The company's new president, Alan Hirschfield, and the head of its telecommunications division, Steve Roberts, foresaw a time when Fox (with its prospective partner, CBS) would become a "video publisher," conceiving productions expressly for release as videocassettes. But first there was a little problem that needed addressing. Fox wanted to establish a two-tiered price structure: inexpensive cassettes for people to buy outright and a more lucrative lease arrangement for the rental side of the business. It had tried to do this by contractual arrangement with the retailers, and the early returns suggested that a lot of them, contract or no contract, weren't cooperating. The only way to enforce such a system, Roberts and Hirschfield had decided, was to persuade Congress to modify the first sale doctrine in order to make the rental of a videocassette a transaction that required the copyright holder's approval. In the Fox view, this would make it possible to impose order on an anarchic market and turn home video into a major element of the movie business. Hirschfield and Roberts had no objection to a royalty on VCRs and blank tapes, but they regarded the rental issue as more important, and before Fox gave its backing to any royalty legislation, they insisted that it address their problem as well.

The other studio executives—Barry Diller and Michael Eisner of Paramount; Ted Ashley and Frank Wells of Warners; Frank Rosenfelt and David Begelman of Metro-Goldwyn-Mayer; Francis Vincent and Frank Price of Columbia; Andy Albeck and James Harvey of United Artists; Donn Tatum of Disney; and Robert Rehme of Avco Embassy—gave Valenti their backing after two weeks of intense negotiations. Among these gentlemen there had been a certain amount of what the loyalty oaths call "mental reservation" about Universal's lawsuit and its prospects. But the Ninth Circuit decision had forced them to upgrade their opinion of Universal's strategic judgment, and when they realized that a relatively modest additional effort of their own might allow them to turn a profit on Universal's and Disney's vast investment of money and goodwill, they found the prospect irresistible.

By the time the Betamax case fell into his lap, Jack Valenti had been a presence in Washington for eighteen years, making him one of the capital's longer-running attractions. At five feet six, silver-haired and square-jawed, he cut a conspicious figure for one who wasn't a movie star or an elected official. Valenti had never cared to hear himself called a "lobbyist," or the MPAA a "trade association," and indeed, those drab terms hardly begin to convey his remarkable place in the taxonomy of Washington personages. The word *ambassador* comes closer. The range of his duties—from escorting visiting dignitaries to hosting parties and ceremonial gatherings to bridging linguistic and cultural differences—is distinctly ambassadorial, as are the ceremonial portions of the MPAA's stately headquarters. As a matter of cold legal fact, however, Valenti is officially registered as a lobbyist, and it is certainly one of his and the MPAA's duties to monitor developments in the legislative and executive branches of government and to represent the movie industry's interests whenever they are at stake. And the belief that Valenti was well qualified to perform that duty was certainly one of the reasons the studios lured him away from his post as special assistant to President Johnson in 1966 and, by 1981, were paying him a salary approaching half a million dollars a year. His political background, in all likelihood, had made more of an impression on Hollywood than his background in the movie business, which consisted of a year as an usher, handyman, janitor, and backup in the concession booth at the Irish Theater in Houston, Texas.

Valenti was the grandson of Sicilian peasants who emigrated in 1881. New Orleans was their point of disembarkation, although Valenti always suspected that they meant to go to New York and got on the wrong boat. Deliberately or accidentally, the family settled in Houston in the 1880s, put down roots, and picked up a twang. Valenti's father was a clerk in the tax assessor's office and a dabbler in real estate. The son, who inherited the volubility and expansiveness for which both Sicilians and Texans are famous, was graduated from high school at fifteen and had early notions of becoming a trial lawyer, like his father's friend Percy Foreman. But after his stint as an usher, Valenti went to work for Humble Oil ("the greatest corporation ever conceived," he always said) and made his way through night school at the University of Houston. He enlisted in the Army Air Corps at the age of twenty, became a pilot, and received the Distinguished Flying Cross for a bombing operation in which his plane, although badly hit, reached and destroyed its target, a ferry system in the Po Valley of northern Italy. After the war Valenti went to Harvard Business School for two years, rejoined Humble in the advertising department, and finally cut loose in 1951 to found an advertising agency in partnership with one of his Humble colleagues.

Soon Valenti was immersed in Texas Democratic politics—so thoroughly immersed that he was summoned to Washington on the same day Lyndon Johnson became president. Valenti had put in a few weeks of logistical work on President John F. Kennedy's Texas trip and was in the motorcade, about seven cars back, when the bullets struck. At Parkland Hospital a White House aide accosted him in a stairwell. "He grabbed me by the elbow," Valenti recalled, "and said, 'Jack, the vice-president wants to see you right now. The president is dead, you know.' This was about ten minutes to one"—Valenti keeps close track of the dates, hours, and numbers in his life—"and I said, 'Oh, my God!' and I started to weep. And he said, 'Get a hold of yourself.' " Valenti was whisked off to Love Field and escorted through three layers of Secret Service men and onto Air Force One for his appointment with Johnson. "He told me that he wanted me on his staff and that I should be prepared to come to Washington with him that day. I said, 'Well, Mr. President, I don't have any clothes.' And he said, 'Well, call your wife and have her send some up.' And I said, 'I don't have a place to stay.' And he said, 'Yes, you do. You gonna live with me.' And I lived in the president's home at The Elms for about ten days and then on the third floor of the White House for about a month and a half until my wife finally came with our new baby and we found a place to live."

His duties in the Johnson administration ranged from speech writing to congressional relations. He did not attract much public attention until 1965, when he concluded a speech to the Advertising Federation of America by saying, "I sleep a little better, a little more confidently each night, because Lyndon Johnson is my president." That line "stuck to me like plaster," Valenti recalled. Johnson was still a popular leader at the time; even so, his powers as a human sedative had not been generally appreciated, and there was some feeling that Valenti had got carried away. He would always insist that the words had been taken out of context. "My text was from Churchill," he would say. He had been recalling what Winston Churchill told the people of France in 1940—"Sleep to gather strength for the morning. For the morning shall come." Some years later Valenti went to see his old boss—now retired—and brought up the surprising endurance of the "sleep a little better" line. Johnson pointed out, consolingly, how rare it was for a presidential assistant to say anything memorable.

A prodigious quoter, Valenti likes to jot down favorite utterances—and their origins—in an alphabetized journal which has grown to three hundred pages over the years. He refuses to use books of quotation, however; *that* strikes him as cheating. He has equally firm ideas about the English language. When the rise of the home video business led some people to label movies a form of "software," Valenti was appalled. "I believe the end result of the creative labor is so fragile, is so full of gossamer lace," he explained to a gathering of the International Tape Association, ". . . that to call it software is an abomination." Nevertheless, he is an ardent phrasemaker who, in the heat of his own oratory, mixes metaphors with gusto.

If the MPAA *were* a mere trade association, Valenti would rate, hands down, as the most widely recognized head of a trade association Washington has ever seen. Few perks of high office are more valued than the invitations that regularly go out to members of Congress and administration officials—and sometimes filter down to their aides—for, say, a dinner honoring Peter Rodino and the congressional Italian-American caucus followed by a screening of *Return of the Jedi*. Few moviegoers have ever lowered themselves into plusher, ampler theater seats than the ones in the MPAA's screening room. Few complain about the quality of the edibles and potables dispensed in the MPAA reception room, whose walls are graced with hand-painted views of colonial Boston, New York, and Philadelphia. And while Valenti would never debase such an affair with talk of something so crude as pending legislation, there are those who contend that these after-hour occasions are the playing field on which the movie industry has won important political victories—in the tax and copyright areas, for example.

During his initial round of calls and meetings Valenti got his member companies to agree to lend the services of their Washington lobbyists—Columbia, MCA, Fox, Warner, and Gulf and Western (the parent company of Paramount) had such people—and to pony up some extra money so the MPAA could retain lobbyists to work expressly on the royalty issue. He also won general support for his choice of a lobbying firm—the firm of Akin, Gump, Strauss, Hauer, and Feld, or, as it is commonly known in Washington, "Bob Strauss's firm."

Valenti's ties to Robert Strauss went back twenty years, to their curiously symmetrical beginnings as ethnic Texans lured away from prosperous business careers by the two great human magnets of the Texas Democratic party: Lyndon Johnson and John Connally. Since moving to Washington to serve as treasurer of the national Democratic party, Strauss had divided his time between his law firm and his political activities, and he had recently completed a tour of duty as Jimmy Carter's special trade representative. Now he was back to more or less full-time lobbying, and doing exceedingly well at it. His annual income had recently been estimated at a million dollars, and the source of some of that money was suggested by a ditty that a group of Akin, Gump lawyers sang to him on the occasion of his sixty-fifth birthday. The tune was one of John Kennedy's favorites, "Camelot." The lyric went:

> We keep all well-heeled oil men out of trouble,
> Protect each cattle baron's precious rump;
> Recession times we only charge you double
> At Akin, Gump.

. . . and so forth. With an extraordinary network of friends in the political and business worlds—"Strauss people" they were called—and an equally extraordinary ability to remember their names, Strauss had a reputation as a consummate behind-the-scenes man. His detractors sometimes intimated that he was longer on connections than substance. His admirers agreed; they just saw the matter in a different light, attributing Strauss's success in law and politics to a refusal to get bogged down in the abstractions and principles that so often led other people to quarrel with one another. "Let's make a deal" might have been his motto.

Oil and gas were the primordial ooze from which Akin, Gump had sprung in the Austin of the late 1940s, and energy interests had sent the firm to Washington in the sixties and seventies. "They were the first Texas law firm that really took Washington seriously," said a lobbyist who has worked with and against them. "They had the best oil and gas people in the city." By 1981 Akin, Gump had the largest Washington contingent of any out-of-town firm—some eighty lawyers in all—and it had signed up nonoil clients such as AT&T, the National Football League, and Gallo wineries.

Strauss's time was, of course, a precious commodity. It was the rare client which could afford to have Strauss handle all its business personally. "Strauss was there whenever we needed his judgment or a key phone call or that sort of thing," Valenti said. For more mundane duties, Strauss designated Joel Jankowsky, an Oklahoman who had been an aide to the late Speaker of the House Carl Albert, and Dan Spiegel, who had been an aide to Hubert Humphrey during his last term in the Senate.

Despite Stan Parris's success with the NFL blackout bill, there was little danger of prompt action in the House, where copyright (for reasons shrouded in bureaucratic history) fell within the province of the Judiciary Committee's subcommittee on courts, civil liberties, and the administration of justice, headed by Robert Kastenmeier of Wisconsin, a congressman known for his deliberate, evenhanded ways. The situation in the Senate was more volatile because its rules, unlike the House's, allowed an unrelated amendment to be added at the last minute to any measure coming up for a vote. So the MPAA turned its attention first to the Senate, and it was forced to go on the defensive before it could get its legislative alternative ready. "Please don't do anything precipitously" was the message the movie industry and its friends in Washington sought to deliver. Ten days after the Ninth Circuit decision David Ladd, the head of the federal Copyright Office (and a strong supporter of the royalty proposal), gave a speech warning of the dangers of haste. "Statesmanship," he said, "is not possible on the fly."

The senator whom the MPAA approached to sponsor its legislation was a natural choice. Charles Mathias, Jr., a liberal Republican from Maryland, was the Senate's leading copyright authority and, as Valenti put it, "a champion of intellectual property, although I don't think it gets him one vote in Maryland." Visitors to Mathias's Senate offices were sometimes treated to an inspection of a musty tome entitled *The History of the Law in Colonial Maryland,* whose chain of ownership Mathias could trace back to the eighteenth century. A man with a high regard for history and literature, he was proud of his modest role in the Nixon impeachment hearings, for that was something, he knew, that would be written about and remembered. Called Mac by one and all, Mathias was known for his moderate views, his cherubic smile, and his affable personality. But when his party took control of the Senate in 1980, it was an ambiguous triumph for Mathias. By the normal workings of the seniority system, he might have looked forward to the chairmanship the Judiciary Committee, but Strom Thurmond of South Carolina had frustrated his hopes by moving over to Judiciary from Armed Services, and as chairman of Judiciary Thurmond proceeded to abolish the antitrust subcommittee, which Mathias was then in line to head, rather than give him so much influence over such an important domain.

Patents, copyrights, and trademarks was another subcommittee that Mathias might have chaired, for he had been its ranking Republican member. But it, too, had been abolished, in 1977, on the theory that the passage of the 1976 copyright act rendered its existence unnecessary. So the authority to hold hearings, call witnesses, and mark up a video royalty bill would rest with Thurmond, as chairman of the full committee, and to the MPAA's regret he let it be known in the early going that he favored the DeConcini bill. "I like to think it was my persuasiveness that did it," DeConcini said later, "but I think there's a Sony plant being built down there, and fortunately they were bright enough to go talk to him. . . ." (Sony had bought land for a TV factory in South Carolina.)

Within a few weeks of the Ninth Circuit's decision, Jankowsky and Spiegel of Akin, Gump and John Giles, the MPAA's legislative expert, had entered into discussions with Mathias and his staff. The proposal they advanced was a simple one. It took the inocent-seeming form of an amendment to the DeConcini bill, tying the idea of a copyright exemption for home videotaping to a royalty the amount of which would be fixed in the legislation itself—at fifty dollars on each VCR and a dollar or two on each blank cassette, if Valenti had his way. But Mathias balked at the idea of naming a figure. Instead, he proposed leaving that task to the Copyright Royalty Tribunal, an agency created by the 1976 act to administer the jukebox and cable TV royalties. With some uneasiness, since there were already rumblings of dissatisfaction with the CRT's performance in the cable area, the movie companies agreed, perhaps hoping that an administrative body would prove at least as generous toward copyright owners as Congress.

The next concession was more traumatic. Mathias introduced his amendment on December 15, and no sooner had he done so than the MPAA heard from an old ally. The movie industry's plans had come to the notice of the recording industry through the medium of Jon Baumgarten, a former general counsel of the Copyright Office. Baumgarten had been retained by the MPAA to draft its royalty legislation, and he was also on retainer to the National Music Publishers' Association. The year 1981 had been a bad one for the recording industry. So had 1980 and 1979. After years of phenomenal growth revenues had flattened out, and the number of new albums produced annually had actually declined by thirty-two percent since 1978. Many people in the business attributed their problems to the use of audiocassette recorders to tape radio broadcasts or records (usually owned by someone other than the taper). The *Wall Street Journal* estimated that taping cost the industry as much as a billion dollars a year.

The record people did not share the widespread belief that the law permitted this activity. But they acknowledged—and deplored—their failure to take legal action when the problem first surfaced. "Why the industry didn't bring a lawsuit, I don't know," Stanley Gortikov, the president of the Recording Industry Association of America, once said. The oversight was probably due, in part, to the low quality of sound delivered by audiocassette recorders in their early days—they didn't seem like much of a threat—and to the industry's preoccupation with other political issues. But the past was past. Now that Hollywood was going after videotaping, the record industry and the societies of composers and songwriters saw an opportunity. If there was to be a royalty on VCR's and videocassettes, they reasoned, there ought to be one on audio recorders and audiocassettes as well, and it should go into the same legislation.

The movie people had no quarrel, basically, with the justice of the record industry's arguments. John Davis, Steve Kroft's cocounsel in the Betamax case, was asked about the audio analogy by a reporter for *Esquire* magazine, and he answered: "What's the difference between us and our claim, and the record industry and a claim against audio recorders? I'll tell you what it is. It's that *they* sat on their hands when they should have sued." Nevertheless, Valenti and his forces were not eager to see the two causes merged. Moviemakers enjoyed special privileges under the law, and they liked being special. Besides, they worried that the ability to record copyrighted music free of charge had become so commonplace and so undisputed that it could no longer be challenged; and they worried that the video royalty would become tainted by association.

People on the movie industry side would say, "Oh, hell, everybody tapes from the radio. This is going to tie us down." But the pressure to let the audio people into the legislation was considerable. The recording industry had its own powerful sponsors in Howard Baker of Tennessee, the Senate majority leader (who had Nashville to think about), and Robert Byrd of West Virginia, the majority-turned-minority leader, who was a budding recording star in his home state, with a record called 'Mountain fiddler'." Baker and Byrd approached Mathias and told him (as a Mathias aide later recalled) that the recording industry "didn't want to be left at the station." If Mathias and Valenti agreed to merge the two goals, Baker and Byrd gave their assurances that the video royalty would gain important support—of which they represented only the advance guard.

At the beginning of March 1982 Mathias introduced his revised amendment, which prescribed royalties for audiotaping as well as videotaping and gave the copyright owner the exclusive right to authorize the rental of any prerecorded tape, audio or video. With Chairman Thurmond's cooperation, Mathias announced that he would chair a series of hearings in April on his proposal as well as DeConcini's.

In the House the MPAA had chosen Don Edwards of California as its point man. Edwards was a liberal of the old school and "a good soldier for the industry," to quote one of his aides, speaking anonymously. (In that, however, Edwards was no different from most of the state's congressional delegation. Over the years the "California Mafia," as it is sometimes called, had represented the wishes of the movie industry with approximately the same zeal shown by the Texas and Oklahoma delegations on behalf of the oil industry, although the aide who volunteered this analogy quickly added that liberal Californians "would cringe at the thought.") On March 3, Edwards introduced what amounted to the same proposal as Mathias's, and Bob Kastenmeier agreed that he, too, would hold hearings on the issue. In the Hollywood camp the panic of the first few weeks receded, and a more relaxed attitude, not utterly devoid of optimism, took hold. "I don't think members of Congress feel it's as urgent as they did," Fritz Attaway, the MPAA's in-house counsel and vice-president, told *Video Week*. "The pressure for a quick fix has subsided now that we've been presenting our side. . . ."

The Hollywood lobbyists began having weekly meetings, sometimes on their own, sometimes with a recording industry contingent there, too. One of their first tasks was to formulate what might be called a "high concept," to use a movie industry term. As Valenti explained, "We had to ask ourselves, 'What is it that we want to say to people? How do we present our arguments?' Lyndon Johnson used to admonish me—and whatever I know about politics I learned from him—to make the congressman a hero for voting with us. How do you make the congressman a hero? You always want to tilt your argument toward how the public might see this as being in their interest. We had to make the case that copyright was in the public interest. We had to make people see that you couldn't charge twenty million dollars for a ticket at the movie theater, so the only way you could make a twenty-million-dollar picture was to collect two dollars or three dollars or whatever every time someone saw it. We had to show that if you allowed people to tape what belonged to somebody else, at some point you'd have to compensate the creators for the losses they were going to suffer in the marketplace. We had to make the argument that in the long run, if people wanted to have a stream of good movies five, six, ten years down the road, the only way to do it was to make sure that people could retrieve their investment and make a little money off of a picture that they'd made."

When they weren't conceptualizing, the proroyalty lobbyists spent a lot of time pooling reports of conversations on the Hill. "Someone would say, 'Well, I talked to so-and-so and he said *this,'* and someone else would say, 'So-and-so said *that,'* " Valenti recalled. "And we'd agree that somebody ought to go talk to this congressman or that senator, who was leaning toward us or away from us. We would parcel out members whom we needed to talk to—particularly on the judiciary committees, but also on other committees." It fell to Joel Jankowsky to keep what Valenti called the "nose-count book," with check marks in a series of vertical columns at the left numbered one through five, to track the status of the senator or representative whose name appeared at the right. "One," Valenti explained, "is a guy who's solidly for you. Two is a fellow who's for you but not solidly. Three is a fellow who's in the middle. Four is some fellow who's leaning against you but not harshly, and five is a guy that you've lost. We'd go after the threes and fours to bring them to us and try to get the twos into the ones. Of course, you always want to make these marks in pencil."

In their daily concourse with members and aides, the agents of the movie industry were pleased to discover that it was often possible to break through an initial reaction of outrage against the Ninth Circuit's decision. It was a matter of "getting people to think about what copyright is—about what the copyright owner of a film has and why it must be protected," according to James Corman, a former congressman who represented the San Fernando Valley for twenty years before his defeat in 1980 and became a lobbyist for MCA early in 1982. "I must say," Corman added, "I never thought about it when *I* was in Congress."

Corman found his ex-colleagues receptive to the argument that "your grocer could sell the bread cheaper if he didn't have to pay the baker." He was careful to point out how little money the individual consumer would have to pay if a royalty were enacted. He reminded his listeners that the TV manufacturers had fought a law requiring the incorporation of UHF tuners on TV sets, claiming that prices would go up. But prices had actually gone down. The VCR business was too competitive for a royalty to have a serious effect on retail prices, Corman argued. The average consumer would probably pay no more than a penny an hour for the use of his VCR, "and if you walked into somebody's living room and told them that they could have all the commercials removed from a program for a penny, they'd be delighted, now, wouldn't they?"

OUTGUNNED

FOUR DAYS AFTER the Ninth Circuit had rendered its decision, William Baker, Sonam's director of corporate communications, had lunch in the Oyster Bar of the Plaza Hotel with an old friend, John Adams, an Englishman who had been a Fleet Street reporter and a TV news writer for ABC and NBC before setting up shop as a public relations man in Washington. Adams was there to impart his wisdom on what, if anything, Sony could do to prod Congress into passing the bills introduced by Senator DeConcini and Congressman Parris. Baker emphasized that Sony, as a Japanese company, would not want to play a prominent role in an American political dispute. That, Adams replied, would be unwise in any case. "One company, especially a foreign company, especially a Japanese company, probably wouldn't get much sympathy if they tried to pursue this alone," he said. "Why don't you form a coalition?"

There are trends in lobbying as in everything else, and the ad hoc coalition was the trend of the seventies and eighties. Ever since General Motors, Chrysler, Ford, and the United Auto Workers had banded together against the Clean Air Act of 1970, strange bedfellows had been pressing for the passage of this piece of legislation or the defeat of that one (which was often the only thing they agreed on). Forming a coalition helped defuse the suspicion that a cause served a special interest at the expense of the general interest, and it helped spread the cost of lobbying—two considerations that would certainly weigh with Sony.

"Well, a coalition of who?" asked Baker. If what Adams meant was a coalition of the VCR manufacturers—all of them Japanese—Baker didn't think they would go for the idea. They might be more favorably disposed, he said, toward a coalition that was broader in scope and not so readily identified with Japan. This got Adams and Baker to thinking about American interests with a stake in the VCR market: consumer organizations, retailers, the chemical companies that played a part in the manufacture of tape, and, above all, the electronics companies that were selling Japanese-made VCRs in the United States.

Later Baker talked the idea over with Sonam's general counsel, Ira Gomberg, and with the company's government relations man, Sadami Wada, and its new president, Kenji Tamiya. Tamiya took the attitude that Sony, as a Japanese company, ought to keep a low profile at all costs. Wada and Gomberg countered that *Sonam* was an American company; it had sixty-five hundred employees—most of them American—whose livelihoods were at stake. After a long-distance consultation with Morita, Tamiya gave his guarded consent. But he and Morita wanted someone outside Sony to take the initiative, so Baker was asked to approach the Electronic Industries Association in Washington, and it was the EIA, with the assistance of John Adams Associates, that organized a preliminary gathering at the Four Seasons Hotel in Georgetown on November 11, 1981.

Before that meeting, Baker, Gomberg, and Wada went up to Capitol Hill, where they discussed the legislative prospects with Parris and DeConcini and their aides. They drew encouragement from the story of Parris's success with his NFL blackout bill, and they liked the idea (contributed by one of the aides) that the DeConcini and Parris proposals sought only a "clarification" rather than a real change in the law. Sony, after all, had claimed in its lawsuit that home taping already *was* fair use. This modest goal would sit better with the people in Tokyo, and besides, Baker recalled, "We said, 'Jeez, a simple clarification—maybe it will fly.' "

The first meeting of the would-be coalition brought together about twenty-five potential participants—most of them lobbyists for the EIA's member companies—around a cluster of tables in a hotel conference room. They started with coffee and doughnuts and wound up with lunch. Wada (who, in keeping with Japanese corporate tradition, used an American nickname, Chris, during his stints in the United States) gave an animated talk on the route the two bills would have to traverse. "It was quite a turn," John Adams recalled. "Here was this Japanese guy jumping up on a chair and drawing pictures for us to show us how Congress worked." The coalition got some further insight from a Democratic congressman, Dan Glickman of Kansas, who gave a luncheon speech in which he endorsed the Parris bill but warned that the House probably wouldn't take any action on it without plenty of outside encouragement.

The assemblage spent a good deal of time debating what to call itself. Jack Wayman, the senior vice-president of the EIA's consumer electronics group, was pushing *The Right-to-Tape Coalition.* But J. Edward Day, a genial old Democrat who had been postmaster general in the Kennedy administration and now represented the EIA as a partner in the law firm of Squires, Sanders, and Dempsey, commented that *Right-to-Tape* sounded dangerously like *Right-to-Life.* Others worried about being mistaken for a movement to get the government off the backs of the nation's wiretappers. Finally, Adams volunteered that the word *home* should appear in the name somewhere since the purpose of the coalition was to lobby for the right to tape TV programs in the home. The name Home Recording Rights Coalition eventually emerged.

The key people at the meeting were half a dozen lawyer/lobbyists—the "Washington reps" for Pfizer, Sears, GE, RCA, Matsushita, and 3M—who came, mostly, to listen, made no commitments, and went away with a collective resolve to lobby their corporate clients for money to launch the coalition in earnest. Over the next several weeks they and the people from Sony and the EIA established a steering committee, elected officers, ordered a letterhead, began having regular meetings to share intelligence and plan strategy, and collected $130,000 in start-up money. Another Washington institution had been born.

The man who emerged as the central figure in the coalition was Jack

Wayman, a forty-nine-year-old trade association man. Like the other Jack at the pinnacle of the MPAA's personnel structure, Wayman had a conspicuous head of hair which had changed color prematurely—to white, in his case, rather than silver gray—and he had the gift of gab. His sentences rarely withstood strict syntactical analysis; but before a listener could ponder anything Wayman said, he was usually saying something else, and his abundant self-confidence tended to carry him over any gaps that grammar or logic refused to bridge.

Washington was Wayman's hometown. He had worked his way through Georgetown University selling TVs. (According to a song composed for a dinner in Wayman's honor, he "started out at Lacy's in appliance, where he turned bait-and-switch into a science.") He ruled over only a part of the EIA—the consumer electronics part—but he had generated millions of dollars for the association by creating and overseeing the Consumer Electronics Show, a semiannual affair (Las Vegas in January, Chicago in July) which had grown into the biggest show of its kind in the country. The winter 1982 edition attracted seventy-five thousand people and filled 603,000 square feet of exhibit space—twenty-two football fields' worth.

Wayman would appear at these extravaganzas in a high-visibility uniform consisting of bright orange blazer and bright yellow pants, and he had more than a little of the circus barker in him. Not being a detail man, he was happy to delegate much of the day-to-day work of running the coalition, and he delegated it mostly to Gary Shapiro, a junior lawyer at Squires, Sanders, and Dempsey, who, later in 1982, went to work for Wayman as in-house counsel. Wayman hired Shapiro in order to cut down on the EIA's legal bills. Shapiro was only twenty-six, and even a three-piece suit wouldn't make him look any older; but he had been a legislative aide to a congressman from Oklahoma, he had done a little lobbying during his time with Day's firm, and he had no pretensions. Better, Wayman figured, to have this personable, well-spoken, smart young lawyer on the premises—and on fixed salary—than off working for somebody who would want $150 an hour for his time. As Shapiro said later, "Jack never wants to spend a lot of money on anything." Shapiro found himself serving as gavel pounder and chef under protest at many of the Home Recorders' meetings. "Jack would stop at a supermarket and get some cheap baked goods and a jug of orange juice and some bacon, and then he would ask me to cook the bacon in the EIA kitchen," Shapiro recalled. "We really spoiled the coalition. They started asking, 'Where's the bacon?' at our meetings long before the 'Where's the beef?' commercials."

At first there was a general desire among the Home Recorders to avoid making too much of the coalition—to have it be a kind of hobby that the founders could manage in their spare time at minimal cost to their clients. They would rely on John Adams Associates (which the coalition retained at Sony's suggestion) to generate press coverage and drum up public support, on Wayman and the EIA staff to handle housekeeping chores, and on J. Edward Day and his firm (already on retainer to the EIA) to keep tabs on the legislative situation. But a majority of the coalition's steering committee soon decided that these resources would not suffice. A big-time lobbyist was needed.

After interviewing a number of candidates, the coalition decided to retain Charles Ferris, who had been a top aide to Mike Mansfield, the former Senate majority leader, and had gone on to work for the speaker of the House, Thomas P. O'Neill, Jr., before spending a term as chairman of the Federal Communications Commission in the Carter administration (in which capacity he had presided over the deregulation of cable TV). Ferris was a partner in the Washington office of a Boston law firm that, with his presence, became known as Mintz, Levin, Cohn, Ferris, Glovsky, and Popeo. The firm evidently held every partner in high esteem, for its receptionists invariably disgorged the whole mouthful of names—"Mntzlvnknfrsglvskpeo"—as their first greeting to callers from the outside world.

Although Ferris was only forty-eight—younger than Wayman *or* Valenti—his hair had turned an autumnal white. It was a curly head of Boston Irish hair over a wide, friendly Boston Irish face. A quick study, he assumed a spokesman's as well as a lobbyist's role. As the home taping issue heated up, TV news and talk shows, consumer groups, and other entities began holding debates and seminars on the subject. Valenti usually represented the movie industry at these affairs. The Home Recorders tended to offer up Ferris to represent their side—which tended to make Wayman furious. He saw no reason why the coalition should pay an outsider a couple of hundred dollars an hour to do what he, Wayman, was already on salary to do—speak for his industry. "People don't want to hear a hired gun," Wayman explained. "I'm the counterpart to Valenti."

The coalition had agreed to pay Ferris fifty-three thousand dollars for the first quarter of 1982 alone, and Wayman was not entirely persuaded of the need for that investment. After twenty years with the EIA and all his adult life in consumer electronics, he just plain didn't trust anyone whose allegiance was up for hire. When a high-priced lobyist signed a client, Wayman suspected, he started thinking about signing the next one. Even when the lobbyist was busy lobbying, there was no guarantee he wouldn't concede a point on Client A's behalf in order to gain a point for Client B.

In the spring of 1982 there came a time when Ferris thought he saw an opportunity to attach the Parris bill to a cable TV bill which was about to be voted out of the Judiciary Committee. "Charlie had about ten hours, as I recall, to develop the language of his amendment and get it to Bill Thomas, a freshman Republican who had agreed to introduce it," one of the Home Recorders recalled. "He got together the best group of legislative minds he could in a hurry and drafted the amendment, and he didn't go through the whole checking process with everybody in the Western hemisphere. And because of problems with the amendment and the fact that we just plain got stomped by Hollywood's muscle, it was ruled nongermane." After this debacle the Home Recorders conducted what one of them later termed a "retribution session," in which Ferris was held to have exceeded his instructions.

Relations between Ferris and Wayman ultimately toned down, with Ferris surrendering most of his spokesman's duties in order to concentrate on lobbying. But tactical issues continued to divide the coalition, and the meetings of its "steering committee" tended to confirm the oxymoronic aspect of that phrase. Fierce arguments erupted between Bill Baker, representing Sonam, and Nancy Buc, representing Matsushita. Baker took a "wild-man approach," according to one member of the coalition's inner circle. "He was always saying, 'Oh, my God, we've got to do something right now!' The lack of subtlety was astonishing." As for Buc, "She's one of those people who can listen to a long, disjointed collection of thoughts on subjects that are completely new to them, close their eyes, and, about thirty seconds later, start to talk in sentences and paragraphs and 'Thus' and 'Consequently' and 'Therefore, what we should do is dot, dot, dot.' That kind of mind I'm very jealous of, and I bow before it. But she has no patience. None."

Buc's and Baker's differences reflected, to some degree, those of their companies. Although Matsushita had made a late start in the VCR field, it had captured a much larger share of the worldwide VCR market than Sony. Thus its stake in the legislative conflict was potentially even greater. But Matsushita was not as agitated as Sony about the possibility of an adverse decision in the Supreme Court or the possibility that the court would simply refuse to hear the case (making the lower court's decision the last word). In part, this was because the Ninth Circuit decision could, in the near term, lead to action against Sony products alone, no other manufacturers having been named in the lawsuit. Even if Universal eventually secured judgments against Sony's competitors, Sony stood to lose a great deal of money in the meantime. With this prospect in mind, Sony took an activist approach to the lobbying. A small team of Sony people took part in the coalition's work, and they continually urged the Home Recorders to be aggressive in their efforts to get Congress to adopt the DeConcini and Parris bills. Matsushita's participation was more modest in scale and more relaxed in outlook. Nancy Buc was its sole representative at most of the steering committee's meetings, and she advocated a defensive strategy aimed, basically, at obstructing the Mathias and Edwards proposals. Apparently content to let the judicial process take its course, Buc argued that it would be impolitic to push Congress too hard. "We fought about it constantly," one lobbyist recalled.

In the early spring an informal survey of sentiment in the House and Senate produced a gloomy reading. Support for the royalty seemed to be increasing, and there no longer seemed to be any strong sentiment for pushing ahead with the DeConcini and Parris bills in their unamended form. An inferiority complex began to infiltrate the coalition. Its members worried that they were in over their heads in a contest with the movie and music industries, which, as one lobbyist put it, "have spent twenty or thirty years currying favor on Capitol Hill, very successfully and honestly, by having people down to Jack Valenti's place for movies, by bringing recording stars in to visit members of Congress, by inviting people to concerts and having movie and recording stars at fund-raisers. They've done a very good job of that," he added, "and given the fact that this is not an 'issue of conscience,' I believe that members are inclined to vote something for them. You'll find that this is especially true in the Senate. Senators tend to want to run for president, and everybody in that capacity has to raise money from Los Angeles or New York. You just can't run for president without getting big bucks from those people."

The example of Senator Edward Kennedy was considered instructive. The Home Recorders saw themselves as defenders of consumers' rights, and they felt that Ted Kennedy, as a liberal, ought to be "with us all the way on this." When Kennedy endorsed the Mathias bill instead, some of the lobbyists perceived a connection between that action and the fact that Lew Wasserman, not long before, had hosted a private party in Los Angeles to raise money for Kennedy's 1982 reelection campaign. Charles Ferris, though, perceived no such connection, and he pointed out that Kennedy had been careful to hear both sides. In fact, he had heard them in the form of a debate staged expressly for his benefit between Ferris and Laurence Tribe, a Harvard Law School professor who had been retained by the MPAA. "We sat on his deck for two and a half hours with Ted Kennedy and two or three of his people," Ferris recalled. "We just went back and forth debating the issue. So Ted Kennedy certainly gave an awful lot of his discretionary time to understanding the complexities of the issue." Later Kennedy talked the question over with a few of his staff members, and "it was a happy coincidence," one of them explained, "that the merits were on the side of a number of people with whom he had had dealings over the years."

The first House hearings on the Betamax issue, on April 12, 13, and 14, 1982, did nothing to restore the Home Recorders' confidence. They were held on enemy territory, in a moot courtroom at the Law School of the University of California at Los Angeles. Just before the first hearing got under way, a delegation of Home Recorders managed to get lost on the UCLA campus and had to ask for guidance from, of all people, Jack Valenti. The fact that the Home Recorders were traveling in a chauffeured limousine (which Wayman had uncharacteristically engaged for the occasion) while Valenti happened by in a rented Mustang was further cause for embarrassment.

Inside the hearing room, too, the MPAA had its act together. Except for Robert Kastenmeier, the subcommittee's chairman, the House members who attended—Don Edwards, Patricia Schroeder of Colorado, and Thomas Railsback of Illinois—were clearly friendly to the royalty idea, and the hearings had been scheduled so that Hollywood's witnesses got the first day of testimony to themselves, with the presence of Clint Eastwood in their number assuring a level of press coverage far beyond the Home Recorders' hopes. In addition to Eastwood—who, in a brief and soft-spoken appearance, communicated his fear that movies would become less attractive as an investment if home taping were not restrained—the MPAA offered a balanced team of witnesses representing theater owners, production people, producers of training films, and other unglamorous parties. When the Home Recorders' turn came, their first two witnesses, Ferris and Wayman, filled up an hour and a half between them, covering much the same ground. When they were finished, Kastenmeier had to caution the remaining antiroyalty witnesses to "summarize and conserve what time remains."

On the third and last day of the hearings the witnesses for the recording industry made a poignant and effective statement of their economic woes, which Valenti cited as dramatic proof of what would happen to the movie industry if videotaping were allowed to reach the proportions of audiotaping. The Home Recorders, preoccupied with the video issue, had not thought to prepare a rebuttal to the audio witnesses.

The Los Angeles hearings prompted another retribution session. The Home Recorders had muffed it. As one of them said later, "We let the other side pick the forum. It was their show. We were sort of along to provide counterpoint, if the media should happen to pick it up." Ten days later Charlton Heston and Beverly Sills were the headliners at a Senate hearing in Washington. "I am not an expert in the law," Heston testified, "nor am I knowledgeable about legislative language and the—to me—mysterious processes from which emerges a congressional bill. Others will better inform you and answer your questions in those areas. But I do know something about filmmaking. I do know something about the high-risk taking that is part of our daily lives. I do know something about the torment and the exhaustion and the long days and nights that go into the creation of a film, from the time the idea is born until that moment of truth when your film begins to unfold in a darkened theater before the ultimate arbiter of its fate—the audience. I know something about all that."

Sills told "a little story that happened to me not more than two weeks ago." A stranger had approached her and boasted of having, on videotape, "the most complete collection of everything I had ever done on television." He had offered to sell her copies of her own performances. Sills asked the senators to "imagine the outcry" if books were subject to unauthorized copying on such a scale or if "operas and concert performances had declined by one-third because of unauthorized admissions. . . . Now, I am not proposing that the freedom of consumers to tape must be denied," she said. "All I am proposing is fair compensation to the creators of that music. . . . Some of my friends here will examine in depth the severe economic impact record duplicating and taping have. But what really worries me is the corrosive effect on the very creative process itself. Record companies will simply be unwilling and unable to take risks on new artists, no matter how brilliantly those beckoning stars shine. A big thumb will be thrust down on creativity."

When Heston and Sills departed, so did most of the TV crews and newspaper reporters, no doubt feeling grateful to the Senate schedulers who had spared them the duty of listening to Ferris, Wayman, and the other low-profile witnesses for the opposition. Several members of the Judiciary Committee also found it necessary to depart at this juncture, in order to attend to pressing business on the floor. "All the senators turned out for Charlton Heston and Beverly Sills," Ira Gomberg of Sonam recalled, "and we got to talk to the legislative aides."

PAPER TIGERS

CONGRESS HAS BEEN through some changes in the last twenty years, and the Washington lobbyist, like any organism faced with an altered ecosphere, has evolved. Ask a young lobbyist what he does for a living, and he will tell you to purge your mind of the bar girls, the late-night liaisons, the envelopes filled with small bills, and the other lurid fantasies perpetrated by Hollywood and the investigative press. Think instead, he will say, of press conferences, hearings, strategy sessions, and memos, of long afternoons at the library and hot nights at the word processor. He will treat you to a history lesson on the decline of the seniority system, the growing tendency to transact congressional business in the open, the proliferation of committees and subcommittees, and the general decentralization of power on Capitol Hill, all of which, he will politely explain, has rendered the old-fashioned lobbyist—the kind who trafficked in favors and connections—obsolete. In the current political environment it would be futile to try to get a law passed by covert means, your instructor will declare, and it follows that *substance,* not *access,* is the modern lobbyist's main stock-in-trade.

Be that as it may, it is probably true that most of what lobbyists do these days is nothing an investigative reporter would wire home about.

The lobbyists employed in the Betamax controversy invested a startling amount of their time in literary labors—in the preparation of memos, white papers, Op Ed pieces, press releases, statistical surveys, scholarly treatises, and "Dear Colleague" letters for the edification of senators, congresspersons, legislative assistants, secetaries, editors, reporters, and other shapers of public policy. Although there are no reliable figures on the tonnage of printed matter generated by the opposing forces in this conflict—that being one of the few aspects of the controversy never submitted to precise statistical analysis—old-timers on Capitol Hill cannot recall a bigger or a more alarming deluge of documents than the one that spilled across their desks and jammed their in boxes in the spring of 1982.

"Unless Congress acts to compensate copyright owners for the home taping of their intellectual property, the audiovisual marketplace will become a barren wasteland of programming that does not edify, nor inspire nor entertain," warned a formidable pamphlet entitled *To Protect and Preserve the American Copyright,* prepared by the law firm of Akin, Gump, Strauss, Hauer, and Feld on behalf of the MPAA and its constituents and allies. With equal conviction, an outline of the issues prepared by the Home Recording Rights Coalition asserted that a royalty on VCRs and blank tapes "would hinder the widespread dissemination of a new technology that promises great benefits for consumers, as well as for the creative community."

From the literature generated by the Copyrightists—as the proroyalty lobbyists may conveniently be called—a reader could learn about the "windfall profits" of the VCR manufacturers. "In 1981, one company alone (Matsushita) had profits which exceeded those of the entire U.S. motion picture industry," one of their handouts reported, while "six out of ten movies fail to recoup their investment and more than fifty percent of the members of entertainment industry-related unions are out of work." *Windfall* was a popular word with the Home Recorders as well. It was their description of what a royalty would mean to the movie industry, and to clarify the financial status of the studios, they helpfully circulated reprints of a *New York Times* article headlined "Hollywood's Hidden Millions: Accounting Rule Masking Future Profits."

If Congress enacted a royalty, the price of VCRs would rise by no more than two percent in 1984 and only about one percent in 1990, because "very substantial portions" of the royalty "would be borne by manufacturers and retailers rather than by consumers," declared an MPAA-sponsored study by the economic analysis firm Robert R. Nathan Associates—a study that came illustrated with almost every known form of chart and graph. But an examination of the same question by the firm of Cornell, Pelcovits, and Brenner Economists, Inc., on behalf of the Home Recorders, concluded that "the consumer ultimately will bear the full cost."

The Copyrightists commissioned a study of the habits of VCR users which found that "approximately 75 percent of VCR households record programs off-the-air for their permanent collections" and "the average VCR household owns almost 20 blank cassettes for the purpose of recording copyrighted programs off-the-air, rather than the one or two blank tapes necessary for 'time-shifting.' " Thus it followed that "building a videotape library is a dominant activity of VCR owners." According to the Home Recorders, though, "Surveys on home VCR usage clearly show that time-shifting of TV programs is the predominant use," while "librarying of television programs is not a significant phenomenon."

Both sides came out with lengthy written analyses of the effects of VCRs on commercial watching. According to the Copyrightists, 87 percent of VCR owners were in the habit of using their pause controls to delete commercials, while 57.4 percent used their fast-forward controls to achieve the same end on playback. Even if, for argument's sake, VCR owners watched every last commercial, the Copyrightists asserted that sponsors and creators would suffer because taping, they said, interfered with assumptions about "program sequencing" and "audience flow." The Home Recorders, for their part, contended that the "data indicate that only a small portion of the VCR audience deletes or skips commercials, and most watch commercials in their entirety when they play back recorded programs."

Both sides published primers on copyright, which explained that the law, as it stood, permitted home taping (according to the Home Recorders) and prohibited it (according to the Copyrightists). And since the value of legal arguments, like the value of clay pots, tends to increase with antiquity, both sides supplemented their findings with scholarly inquiries into the constitutional aspects of the controversy. It seemed that the Founding Fathers, even as they had been forming a more perfect union, establishing justice, insuring domestic tranquillity, and providing for the common defense, had found time to address the issue of home taping, albeit indirectly. Laurence Tribe, the Harvard authority hired by the MPAA, prepared a forty-nine-page treatise with 125 footnotes, which conclusively showed that a law allowing uncompensated and unauthorized taping would violate the Fifth Amendment's clause on taking property without due process and just compensation. "What congress proposes to take without compensation is not only property," Professor Tribe wrote, "but liberty as well, because copyright is a special kind of property—a right to possess, and to control the reach of, one's own expressions." A videotaping exemption, he added, would "endanger, and might indeed impermissibly abridge, First Amendment rights" because "motion picture and television producers will speak less often if the reward for their efforts is greatly reduced."

The Home Recorders' reply, composed by two Washington lawyers, Robert Bruce and Ricki Rhodarmer Tigert, was only half as long (twenty-five pages) with only half as many footnotes (sixty-nine), but it was equally decisive. "VCR usage," Bruce and Tigert wrote, "enhances the exercise of First Amendment rights by expanding the availability of a scarce broadcast resource." The Mathias and Edwards bills, they warned, threatened to trespass on the privacy rights guaranteed by the Fourth Amendment because the process of setting and distributing a royalty might well require an intrusive examination of VCR users' viewing habits. "One particularly frightening possible remedy," they went on to observe (omitting to mention that it was a remedy no one had seriously proposed), "would be the installation of a device on every VCR to monitor everything that is recorded, in order to set charges for recording of copyrighted material." Even the alternative of using random surveys, which "might seem relatively innocuous," was worrisome because the role of the government "would add an important coercive pressure on individuals to participate."

Writing is widely regarded as one of the higher forms of human expression, so the modern's lobbyist's commitment to the written word can be interpreted as a sign that he is evolving in an upward direction. A parallel development which has been cited in support of this theory is the emergence of grass-roots lobbying as a major tool of the profession. The old-fashioned lobbyist was part of the insulation separating the electorate from Congress. The grass-roots lobbyist is a conductor of energy from the former to the latter. As a rule, he goes about his task by identifying people with a clear reason to support a particular industry's viewpoint and then getting them to send (or, at any rate, to sign) letters and petitions to their elected officials. But there are times—and the Betamax case, seen from the movie industry's vantage point, was one—when the process needs to be preceded by a campaign of public education. As Jack Valenti explained, "If you're working for the banking industry or the real estate people, or cable, or broadcast TV, or hardware or fruit and vegetables or you name it, you've got grass roots all over the place. If somebody wants to take away expense account lunches, every restaurant owner in America is rearing up on his haunches. Our grass roots are in two states, New York and California. From a political standpoint we're paraplegics."

In January 1982 the MPAA engaged the public relations firm of Wexler and Associates to address this problem. The firm had been founded after the 1980 election by Anne Wexler, a White House Aide in the Carter administration. But it was already a runaway success, and Wexler was about to broaden her base, ideologically, by bringing in Nancy Reynolds, who had been the press secretary to Ronald Reagan in his California days. At Wexler and Associates, the MPAA's problems were placed in the commodious lap of Dale Snape, a cheerful, bearded former official of the Office of Management and Budget, who, going the conventional grass-roots route, began recruiting production companies, organizations of writers, directors, actors, and craftspeople, and anyone else he could get, to lend, if nothing else, their names to the campaign for a royalty.

The Hollywood forces also decided to run an experiment in direct mail—an exceedingly expensive undertaking—and to retain a firm called Targeted Communications for that purpose. A year earlier Targeted Communications and its founder, Rob Smith, had made a strong impression on the lobbying community with a mailing sent out on behalf of independent telephone companies with an interest in the pending breakup of AT&T. A few people on Capitol Hill had complained about the fact that the envelopes containing the anti-AT&T appeal (which concerned the consequences of certain actions by the Justice Department, if unthwarted by Congress) resembled phone bills and carried the provocative message "Telephone Rate Increase Enclosed." What struck lobbyists about that mailing, however, was the extraordinary rate of response it had provoked. Roughly twenty percent of the recipients had felt moved to communicate with their legislators, and many of them, unlike the usual run of congressional correspondents, were ordinary members of the vote-casting public—people without direct ties to the affected companies.

Targeted Communications' specialty was building "ad hoc constituencies" for causes that lacked ready-made popular appeal. After consultations with Snape and the MPAA, Smith reached the conclusion that older, relatively affluent citizens—especially those who enjoyed the benefits of cable TV and did not own VCRs—would make the best target. Homing in on six states represented on the Senate Judiciary Committee and on sections of those states with the proper demographic characteristics and a high proportion of cable TV subscribers, he culled sixty thousand names from various mailing lists in the company's computer bank. Then he did his best to draft a letter with the right blend of seriousness (it was four dense pages long) and desperation ("If you're going to do it right, you've have to make the letter scream at you," Smith explained), and he invested it with the full personalizing powers of the word processor and the laser printer.

The envelope identified the sender as Charlton Heston, gave a Hollywood address, and offered no other clue to the contents. Inside, the recipient was informed that "what you do in the next five minutes could well determine whether you will see movies like these classics *ever again:*

"*. . . Gone with the Wind*
"*. . . The Godfather*
"*. . . To Kill a Mockingbird*
"*. . . The Philadelphia Story*

> The American movie industry [the letter continued] is under attack. And the quality and quantity of movies that will be available to cable and network television could change dramatically in the years ahead.

> I am writing you today to alert you to this danger and to ask for your help.
>
> Over the past few years, a group of wealthy, powerful Japanese electronics firms have invaded the U.S. video recording market—trampling U.S. copyright laws and threatening one of America's most unique and creative industries. A U.S. appeals court has ruled that the manufacturers of these machines have violated our copyright laws. . . .
>
> The VCR manufacturers have grossed enormous profits in this country, but they have refused to fairly compensate the writers and producers of movies for the use of their copyrights. . . . And now they want Congress to overturn the appeals court by granting an exemption to America's copyright laws without fair compensation.
>
> The result:
>
> Movies might be limited in number to only the "formula movies" guaranteed to make a profit in the theater;
>
> And the best of the remaining movies might be delayed a year or two or eliminated altogether from cable or network T.V. showing. . . .
>
> Only the action of thousands of knowledgeable, concerned Americans like yourself who care enough to fight for fairness will protect the U.S. movie industry, thousands of jobs and the American consumer.
>
> Won't you take the time to urge your Senator to vote for the Home Recording Act of 1982 as embodied in Amendment 1333 of Senate Bill 1758.

The reader was told that "it would be best if you could write your own letter, in your own words." In case he couldn't, a "sample letter for your convenience" was enclosed on the recipient's simulated letterhead. ("We must not let the foreign VCR manufacturers severely damage the American movie industry by running roughshod over U.S. copyright laws," the letter within the letter declared. "We cannot afford to lose another important American industry.") And in case the recipient didn't feel like a trip to the mailbox, a toll-free number was provided so that he could "authorize us to send, at no cost to you, a public service message on your behalf to your Senator."

At Targeted Communications the Heston mailing was hailed as a success. Rob Smith estimated that ten percent of the recipients were spurred to act, and though that was only half the response rate of the anti-AT&T letter, it was an impressive figure in a field where two or three percent response had been the rule. The verdict at Wexler and Associates and the MPAA was less upbeat. Congressional staffers tend to separate mail into two categories—spontaneous and generated—and "we got some anecdotal evidence," Dale Snape said later, "that they treated it as generated mail." The fact that the return letters were printed in an assortment of pastel colors suggestive of decorator tissue paper may have contributed to that feeling; it was certainly the cause of some amusement on the Hill. As Snape said afterward, "It didn't take a real genius to figure out that it was organized."

A year before he began working for the Electronic Industries Association and the Home Recording Rights Coalition, Gary Shapiro had coauthored an article for the *Legal Times of Washington* entitled "Lobbying: Alternatives to Arm-Twisting or Paper Churning." Shapiro and his collaborator had recommended "tapping constituent groups and bringing them to bear on lawmakers in Congress." Letters, petitions, and phone calls could be helpful, they wrote, but "a meeting with the member and/or his staff is essential," and "at all times during the meeting, the attorney must stress the relationship between the client's interest and the member's constituents. Objective arguments, however solid, lose appeal if this connection is not developed. . . ."

One day in December 1981, when Shapiro was starting to work on the Betamax affair as a lawyer representing the EIA, he found himself reading the video royalty amendment introduced by Senator Mathias. He was struck by some language he found on the last two of its twenty-one pages—"buried in there," he said later. After a long discussion of the royalty and the procedures for setting the amount and disbursing the proceeds came a sentence stating that no one, without the permission of the copyright owner, could dispose of a copy of a video recording "by rental, lease, or lending, for purposes of direct or indirect commercial advantage."

This was the provision for which Twentieth Century-Fox—unable to control the rental business and unsatisfied with its share of the take—had successfully lobbied. Shapiro knew nothing of that history, however. He wasn't even sure what the words meant. But he sensed that they would be of interest to the thousands of video rental stores which had sprung up around the country in the preceding few years—and he sensed correctly. He put in a phone call to a man he didn't know, Weston Nishimura, in Seattle. He had read in a magazine article that Nishimura owned a chain of video stores and was (like Rocco LaCapria in New York, and others) trying to establish a national organization of video retailers. "There's this thing in there that I think will affect you guys," Shapiro said after he had introduced himself. He read the section aloud. Nishimura explained its practical consequences—it would give the movie industry the power to demand a share of the proceeds of every videocassette rental or to bar rental altogether if it chose to—and he confirmed Shapiro's suspicion that video dealers liked the law just the way it was. Shapiro suggested, and Nishimura agreed, that it might be in the video dealers' and the Home Recorders' interests to form an alliance.

In January Shapiro flew out to Las Vegas for the Consumer Electronics Show, and he set up a booth to gather signatures on "Defend the Right to Tape" petitions and to dispense "Defend the Right to Tape" T-shirts and buttons. Hundreds of video dealers were at the show, and when Shapiro sounded them out on the Mathias amendment and its rental clause, he was pleased to discover that they became incensed. It helped that many of them were already incensed by the studios' attempts to market their videocassettes on a lease rather than a purchase basis. The dealers might be a little out of control at the moment, Shapiro said to himself as he observed them on the warpath against Warner Brothers. But here, it seemed to him, was a force worth harnessing.

Shapiro and Jack Wayman, the head of the EIA's consumer electronics group, returned to Washington as strong partisans of an alliance with the video dealers. But the idea got a mixed response from the Home Recorders. The lobbyists for the electronics companies were uneasy about an issue that involved software rather than hardware. The tape companies had close ties to the movie industry—they supplied materials for prerecorded cassettes as well as blank ones—and some of their representatives felt that a one-front war with Hollywood was sufficient. There was a certain amount of enthusiasm in the coalition for using the rental issue as a bargaining chip—a position to be abandoned in the clutch, in return for the MPAA's abandonment of the royalty. But others were angered by that idea, and one of the company reps gave a lecture along these lines: "You don't last in this town if you don't play fair, and this ain't playing fair."

Without a mandate—and financing—from the Home Recorders, there was little that Shapiro could do to test his faith in the value of the video dealers as allies. But in a quiet way he pursued his efforts to mold them into a political force. To begin with, he took it on himself to bring together two rival groups of dealers—the Video Software Dealers Association (Nishimura's group) and the Video Retailers Association (LaCapria's). The VRA had more members, but they tended to be militant, small-scale operators—the kind that had threatened to burn down the Warner booth in Las Vegas. The VSDA's members were "the highfalutin video retailers—the ones with the ties," according to Shapiro, and he regarded them as the more presentable from a political standpoint. The VSDA also had a financial advantage, since the National Retail Merchants Association had contributed a hundred thousand dollars, along with office space and personnel, to help it get organized. At the summer Consumer Electronics Show, in Chicago, Shapiro persuaded the leaders of the VSDA and VRA to hold peace talks with himself as mediator, and they agreed to merge under the VSDA banner. Now, at least, if the Home Recorders ever wanted any help from the video dealers, they would know where to call.

Sony had done its best to put an American face on the anti-Hollywood lobbying effort by seeking out allies like Sears, which sold VCRs, and 3M, which made videotape. But the work of the Home Recorders was closely monitored by two high Sony officials from Tokyo: Vice-president Sadami ("Chris") Wada and Deputy General Counsel Teruo ("Ted") Masaki. Often they were the only Japanese in attendance at coalition meetings, congressional hearings, and appointments with members of Congress and their aides. Wada and Masaki didn't always have a lot to say on these occasions, but they were eager to learn. "The term *lobbyist* is very foreign to most Japanese," Masaki explained. "I don't think we have an equivalent. You hear about them, you read about them, but until you actually meet them and see them function, you find it very difficult to envision what lobbyists are. In Japan it is more normal to see the administration introducing bills into the Diet after having worked out the content of the bill through the various administrative agencies and so forth. You seldom see a legislative measure being introduced by our Diet members. I think that's where the basic difference lies."

Sony had been reluctant to play the lobbying game. Morita's willingness to allocate even a limited amount of money to the legislative struggle was a reflection of the extreme urgency of the Ninth Circuit decision as he saw it. "We did not anticipate that kind of a decision; it was a moment that was very grave and shocking," Masaki recalled. VCRs had become the dominant factor in Sony's finances, accounting for almost a third of the company's gross sales and about half its profits. So the prospect of a law that would make them more expensive was viewed with alarm. Pride was also at stake. Sony had been portrayed as a low, parasitical form of commercial life. Jack Valenti liked to refer to VCRs as "millions of little tapeworms" which (he seemed to suggest) were eating away at the core of the American movie industry. The lawsuit and the Ninth Circuit decision implied that the value of a Betamax lay not in Sony's technical ingenuity but in the artistic ingenuity of others. A recording made on a Betamax was, in a sense, another kind of "cheap copy"—the bad rap that Japanese exports had labored under for years.

After a few months in and out of Washington, Masaki and Wada formed the belief—and duly communicated it to Tokyo—that their team, the Home Recorders, was out of its league. Sony's response was to appeal to its competitors for help. Among the Japanese electronics companies, only Sony and Matsushita had become involved in the struggle so far, and Matsushita's attitude struck Sony officials as complacent. Hitachi, Sharp, Mitsubishi, Akai, Toshiba, Sanyo, and NEC all were actively selling VCRs in the United States. A royalty or an injunction threatened those companies, too, Sony pointed out; if they wanted the VCR market to prosper, they ought to foot some of the bill.

In Japan Sony had long been regarded as an aggressive, difficult company, and it did not have the smoothest relations with the rest of the electronics industry. Rather than take Sony's word for how things were going, therefore, the other companies asked the Electronics Industry Association of Japan to obtain an independent reading of the situation. The EIAJ, in turn, put the question to its Washington representative, William Tanaka, a California-born lawyer who had learned Japanese in an internment camp during the war. And Tanaka assigned a young lawyer from his firm, Robert Schwartz, to attend the Home Recorders' meetings and file reports to be forwarded to the EIAJ in Tokyo. The drift of these reports, according to one lobbyist who saw them, was that the Home Recorders "were getting their head handed to them."

At Tanaka's suggestion, the EIAJ decided to hire two additional lobbyists to work on the issue. One was Ronald Brown, a former chief counsel to the Senate Judiciary Committee and, significantly, a partner in Patton, Boggs, and Blow, perhaps the most successful and prestigious firm of lawyer/lobbyists in Washington. Known as "Tommy Boggs's firm" (just as Akin, Gump was known as "Bob Strauss's firm"), Patton, Boggs, and Blow represented banks, oil, sugar, and shipbuilding interests, and it had recently been of some assistance in helping another client, the Chrysler Corporation, obtain a one-and-a-half billion-dollar federal loan guarantee. The firm had branch offices in London and Al Khobar, Saudi Arabia. Tommy Boggs himself was the son of two members of Congress from Louisiana—the late Thomas Hale Boggs, Sr., who had died in a plane crash in 1972, and Lindy Boggs, who had taken her husband's place. His familiarity with the House (and its with him) went back more than three decades, to a time when he had been bounced on his father's knee at meetings of the Ways and Means Committee. Boggs's firm, like Strauss's, had its own political action committee, or PAC. On the personal level, too, Boggs was an active campaign contributor and one of the few men in Washington who could match Strauss, connection for connection.

Tanaka also called the EIAJ's attention to the talents of David Rubenstein, a former domestic policy aide in the Carter White House who had become a partner in the law firm of Shaw, Pittman, Potts, and Trowbridge. A political consultant who has worked with Rubenstein described him as "a very smart, terribly serious technician—typical of the Carter White House domestic policy office." Bespectacled and bookish-looking, Rubenstein was, like Shapiro, a grass-roots aficionado who believed that constituent pressures, properly applied, carried more weight with Congress than private influence or campaign contributions. Rubenstein was asked to head up a grass-roots campaign against the royalty, and in the summer and fall of 1982 he made several trips around the country explaining the importance of the issue to electronics companies and video stores and helping them organize letter-writing campaigns. The alliance that Shapiro had been advocating since his trip to Las Vegas was now, de facto, a reality, and the video dealers proved to be an especially vigorous grass-roots constituency. In their petitions to Congress they described VCRs as "bonanzas" for the movie companies and denounced the royalty as "double payment." They warned of huge price increases and the "ruin" of thousands of retailers if the movie industry succeeded in its quest for legal control over videocassette rentals. "Needless to say," one of their other communiqués concluded, "there are more people in this area who enjoy renting movies and using recorders than there are studio executives."

With Rubenstein, Brown, and Patton, Boggs, and Blow in the ring along with Ferris, Valenti, and Akin, Gump, the home videotaping controversy became a heavyweight contest—and one that, it soon became apparent, would not be settled in an early round. On June 14, 1982, the Supreme Court granted certiorari in the Betamax case. The MPAA and its allies had been hoping the Court would refuse to hear Sony's appeal, creating a legal emergency which Congress would have to confront. "We were sort of gung ho," Dale Snape recalled. "We were thinking, 'We'll get the legislation—we'll solve both problems simultaneously.' " After the Court's announcement the Copyrightists issued a statement vowing to press forward on the legislative front regardless of what happened on the judicial front. But being realists, they understood that a chance to do nothing and blame it on another branch of government would be hard for Congress to resist.

VIGIL

THE INTERIOR OF the Supreme Court could hardly have any more columns if Cecil B. De Mille had designed it, and dozens of reporters had to settle for column-obstructed views of the oral argument in *Universal* v. *Sony,* which followed the justices' lunch break on the afternoon of January 11, 1983. "It hasn't been this crowded since the Akron abortion case," one old Court hand told a newcomer from the video trade press. Even the abortion issue had not provoked the number of amicus curiae briefs—twenty-seven—filed on Sony's and Universal's behalf. The Court had never had so many friends before.

Dean Dunlavey, representing Sony, had a record of two wins and no losses in his Supreme Court appearances. His two previous encounters with the Court could almost have been one, however. Both involved conflicts between federal and state laws governing the advertised weight of foodstuffs—bacon, in the first case, and flour, in the second. In both instances he had persuaded the Court to apply a federal standard allowing the seller to cite the weight at the time of packing rather than at the time of ultimate sale (when a small amount of liquid had drained away or evaporated). Perhaps the memory of these triumphs was a source of confidence; perhaps he drew on some other, more natural well. In any case, as Dunlavey began to address the Court, he seemed unaffected by the setting or the knowledge that six years of effort and millions of dollars of expense might be riding on his performance in the coming half hour. The only emotion that showed in his delivery was contempt for the Ninth Circuit, which, he told the justices, had "rolled over" the facts of the case and "substituted its own impressions in every respect" as though "there had been a whole new trial in absentia. . . ." Universal and Disney were "dead serious" in their desire to remove the Betamax from the market, he said. The "ultimate issue," according to Dunlavey, was "whether under the current law Americans are going to be denied the benefit of time shift home television viewing because a few program owners object."

When Dunlavey pointed out that every Betamax customer received a printed warning against copyright infringement, Justice William Rehnquist asked if it wasn't true that "the package was delivered to the purchaser with that thing wrapped up inside so he'd never see it until after he got it home." Rehnquist also challenged Dunlavey's assertion that Universal and Disney had "put these programs on television intentionally with the purpose that anybody who has the means of doing it can receive it." That was like saying that a publisher who sends his books to a bookstore "should have no objection to people making a lot of Xerox copies of the book."

"It's more like the publisher . . . who, having put his book in the bookstore, knows that whoever buys it can then read it, give it to a friend," Dunlavey countered.

Justice Byron White defended the quality of the Ninth Circuit's opinion against Dunlavey's harsh evaluation of it, and he questioned the claim that Universal and Disney represented the views of a small fraction of copyright owners. Many television shows, White noted, contain musical compositions whose copyright owners might object to the taping of their works, regardless of the sentiments of the owners of the surrounding program. An amicus brief citing just that concern had been filed by the National Music Publishers' Association, he added.

"Those people have come out of the weeds in just the last couple of weeks," Dunlavey said. But he conceded that the "copyright within the copyright" was a "problem."

Stephen Kroft seemed more conscious of the occasion—and more temperate. He tried to soften the Court's perception of his clients and their aims. He spoke of a royalty as one of "several alternative remedies that are available and, I think, are even likely. . . ." The plaintiffs were not suggesting that "an individual Betamax owner in his home is the same as a commercial pirate," Kroft said. ". . . We're talking about millions of Betamax owners, and when they get done making their copies, they end up with millions of copies. . . . It just makes no difference," he argued, "that the millions of copies end up in their hands because they make them individually, without seeing or hearing each other, rather than buying them from a film pirate. The end result is exactly the same."

Kroft's presentation was "on a loftier plane than Dunlavey's and showed flashes of eloquence," in the opinion of Jim Mann, the Supreme Court correspondent of the *Los Angeles Times,* who evaluated the arguments in a column for the *American Lawyer.* But neither attorney had come off all that well, Mann concluded. "Listening to the Betamax arguments," he wrote, "one had the sense that technology is hopelessly outpacing law—and, for that matter, the English language. . . . All too often, both lawyers appeared overwhelmed. They looked like a couple of linguists assigned to explain the computer in a half hour to a panel of ancient Greeks."

Dunlavey didn't like the way the argument had gone either. In his mind, though, the Supreme Court was the problem. The way he saw it, he should have been able to begin by saying, "You know the facts. I won't repeat them. Here's my argument." Instead, he said later, "I'm forced to burn up half my allowable time just laying out a factual groundwork because I know there are judges up there who have not read my brief. There are judges who will figure, 'We have briefs in this case that are probably six inches tall if you stack 'em together with all the *amicus* briefs, and that's kind of a nuisance, so let's four of us get together and pick one of our clerks and let that one clerk read all the briefs on both sides and give us a memorandum on what's in there.' Now I don't give a goddamn if the clerk spends the rest of his life on the brief. He's not the guy I'm trying to reach. I write every sentence in there with the expectation and hope that those nine judges will read it. They have been put there because somebody has said they've got the correct mix of intelligence and education and experience to be passing on these ultimate questions that concern the whole nation. If they're there for that reason, why in the hell should they give those clerks who are one or two years out of law school so much authority? When the clerks come to the law firms, their heads are as big as watermelons. It takes them a year to get through their clerkship and the rest of their life to get over it. A clerk may not pick up a point that I would be sure an older, more mature judge would have picked up. And some judges may not have even read their clerks' memorandum. I've got to assume, as I start off, that there are judges who are thinking, 'I wonder what a Betamax is.' And when I'm in the middle of my factual presentation, I've got to expect that some judge like Rehnquist—who is as sharp as a tack and will have read all the briefs—will say, 'Counsel, here on page thirty-five of the transcript some witness said this or that,' and give me a question that hits right at the heart of my ultimate argument, so I have to be prepared to answer him and then come back to my kindergarten presentation for the rest of them."

A few months after the argument the lobbyists began to mount a vigil at the Court, like nervous nobles awaiting the birth of a monarch. On Fridays they would call a special phone number for a recorded announcement of the day or days in the following week when the Court would be issuing decisions, and on the appointed days—usually Mondays—a line would form outside the public information office beginning at nine in the morning, on the chance that an opinion in the Betamax case would be among the sheaf of papers dispensed at ten. "It was wonderful," Toni House, the court's public information officer (and a former reporter for the *Washington Star*), recalled. "It was a self-policing line. They all knew each other. One lady had a campstool that she brought to sit on." Court employees joked about the high-priced talent that served time on the line—including, on at least one occasion, the former postmaster general of the United States, J. Edward Day. Who could care so deeply, they wondered, about the half hour that might be lost if the lobbyists inquired by phone and withheld their messengers until they knew the moment had arrived? From the lobbyists' perspective, however, every minute was crucial. It would be their responsibility to put the proper spin on the decision as it sailed out toward the nation's living rooms. They would have to study it, arrive at the most favorable reading of its import and ambiguities, and translate that reading into language suitable for public consumption. And all this would have to be done by early afternoon if Valenti, Wayman, Ferris, and the other spokespeople wanted to be guaranteed a prominent place in the evening news shows and the next day's newspapers. Whatever the outcome, both sides anticipated an immediate resumption of legislative hostilities, and they intended to get off to a running start.

It would fall to Dale Snape of Wexler-Reynolds to handle the logistics of bringing a selection of the nation's press corps and its cameras and microphones into a room with Jack Valenti for the official announcement of the movie industry's reaction. This proved to be an exhausting responsibility. When it became necessary for Valenti to be away from Washington on potential D days, it became necessary for Snape to organize a contingency plan. He would reserve a hotel conference room in whatever city Valenti happened to be passing through. "I got to know the people at the Beverly Hills Hotel very well," Snape recalled. Finally, the pressure became too much for him, and he put in a desperate phone call to Toni House at the Supreme Court. "Toni, *please,*" he said, "I don't want to know when the decision is. I understand that's against the rules. But we have these people standing up there every decision day waiting for a decision. Can't we figure out a way where if it's *not* going to happen over a period of time you could let us know it's not going to happen, so we wouldn't have to do this?" Rules were rules, however. They couldn't figure out a way.

Toward the end of June, as the pace of judicial business quickened in anticipation of summer recess, the number of potential D days grew to three and four a week, and with only a few cases remaining undecided, the passage of each likely day focused attention on the next. Relief—of a sort—came on the last day of the session, July 6. "Dozens of lawyers, law firm employees and messengers," the *Washington Post* reported, were lined up outside the public information office when word came down that the case had been "restored to the calendar for reargument." The justices had decided not to decide. They would have the lawyers back again in the fall. In recent years the Supreme Court had been postponing a few decisions in this way at the end of each session. Usually, though, it had identified some unresolved issue which the lawyers were asked to explore further on the next go-round. With *Universal* v. *Sony* the Court exercised one of the exclusive powers of its supremacy: the power to make a ruling without giving a reason. And nobody had a better explanation of its behavior than Dean Dunlavey, who said simply, "They couldn't get five justices to agree on any one decision."

PIGS VERSUS PIGS

WITH THE BETAMAX case safely tucked away in the judicial bedchamber, a less devoted group of lobbyists might have seen a chance to rest from their labors. But the Home Recorders and the Coyprightists carried on with exemplary vigor. Jack Valenti, Charles Ferris, and Jack Wayman kept on debating, before the congressional arts caucus, the National Association of Attorneys General, and the International Television and Video Association; on the Cable News Network, the Larry King radio show, and the "CBS Morning News." Both sides generated leaflets and Mailgrams at the same prodigious pace, and they continued to hold their weekly meetings, although Gary Shapiro, fed up with making bacon for the Home Recorders' breakfast assemblies, persuaded them to convene later in the day, when he could get by with a buffet of M&M's, fruit, cheese, potato chips, and soft drinks.

The hearings, too, went on much as before, with some of the witnesses gradually committing their testimony to memory. One fairly regular participant, Stanley Gortikov, the president of the Recording Industry Association of America, could be counted on to take a blank audiocassette out of his pocket and pull it apart, explaining, "Here is what the home tapers use to harm our industry. It is nothing but a pile of plastic and chemicals and oxides and spindles, and it is useless until it comes alive with copyrighted music from those who we represent. It is just plastic—and that hurts." Jack Valenti was another witness who liked to reach into his pocket—in order to produce a letter from an advertising executive of Frito-Lay, warning of that company's extreme reluctance to pay the networks' ad rates if VCR ownership continued to grow, or a magazine advertisement for a product called the Killer which deleted commercials during the process of recording a movie, as long as the commercials were in color and the movie was in black and white. The hour was near, Valenti would add—although the Home Recorders scoffed at this prediction—when a similar device would be developed for all programming and when it would be built into the VCRs themselves. In the meantime, of course, remote controls with pause buttons existed to fill the same need. Valenti generally came armed with one of these, too, and as he waved it in the air for emphasis, he would cite the example of his own son, who "sits there and pauses his machine and when he is finished with it he has a marvelous Clint Eastwood movie and there is no sign of a commercial." On one occasion Valenti gravely testified that his son hadn't "seen a commercial in eighteen months." To which Charles Ferris replied: "Teenagers are very creative in the ways that they rebel against their parents, and I can think of no more creative way for your son to rebel against you, Jack, than to zap out the commercials on his VCR."

At the outset of each hearing the lobbyists would deposit heaping stacks of statements, studies, and press releases on the press table, and the reporters would glance at them just long enough to assure themselves that they contained nothing new. "I've listened to these people so many times, I could write this story right now," one reporter remarked as a Senate hearing—the fourth on the Betamax issues—got under way in October 1983.

As the press grew weary of the issues, it turned its attention to subsidiary matters, such as the astonishing number of prominent former officials of the United States government who had gone to work for one side or the other. In the early stages of the controversy the Home Recorders and their Japanese constituents had been the more eager headhunters. Charles Ferris had persuaded the coalition to hire two former associates from his days as chairman of the FCC: Nina Cornell, who signed on as an economic consultant (and prepared the Home Recorders' analysis of the effects of home taping on TV viewing), and Robert Bruce, who became Sony's regular Washington lobbyist (and coauthored the coalition's retort to Professor Tribe's essay on the constitutional aspects of the issue). Ferris was also responsible for getting the coalition to retain Marlow Cook, a former Republican senator from Kentucky, on the ground that Cook's contacts among Republicans would neatly complement Ferris's among Democrats. And it was at Ferris's and Nancy Buc's urging that the coalition engaged Carol Tucker Foreman, a consumer activist and a former assistant secretary of agriculture, to mobilize consumer groups against the royalty.

On the same side of the struggle the Electronics Industry Association of Japan had engaged Ronald Brown and David Rubenstein (previously with the Senate Judiciary Committee and the Carter White House, respectively), while a video store group that called itself the Committee Against Regulation of Video Enterprises had hired Rubenstein's old White House superior, Stuart Eizenstat.

The MPAA, for its part, had mobilized a fairly lean army at first, with Valenti, Strauss, and Anne Wexler as the big guns. Thanks to the largess of the recording industry, however, the proroyalty forces also commanded the services of Dean Burch, Richard Nixon's chairman of the Federal Communications Commission; Lloyd Cutler, Jimmy Carter's White House counsel; and Alan Greenspan, an economic adviser to Presidents Nixon and Ford (and a man who, as Charles Mathias remarked at one of the Senate hearings, "has probably testified more often before Congress than any other living human being"). Toward the end of 1982 the movie studios went on a hiring binge. When Thomas Railsback, a Republican member of the House subcommittee with jurisdiction over copyright matters, went down to defeat in the 1982 elections, he was promptly invited to join the MPAA as executive vice-president. When Bruce Lehman, the chief counsel to the same subcommittee, decided to leave his job, he, too, had no trouble finding a place in the private sector; he entered the Georgetown law firm of Swidler, Berlin, and Strelow, with Twentieth Century-Fox as a principal client. And Warner Communications found a spot for Lehman's good friend and former subcommittee colleague Timothy Boggs (no relation to Tommy, who had been retained by the other side). A naturalist, commissioned to do a study of the congressional aide as a class of Washington organism, might have been tempted to classify him as a larval form in the life cycle of the lobbyist.

Deborah Leavy, who joined the staff of the Kastenmeier subcommittee when Lehman and Boggs left it, remarked that it was becoming harder and harder to find a "copyright virgin"—an expert who had yet to go on the payroll of some interested party. At the first Senate hearing on the Betamax issue Joseph Waz, the deputy director of the National Citizens Committee for Broadcasting, a public interest group founded by Ralph Nader, endorsed the idea of a copyright exemption for home taping but proposed a congressional study to determine if copyright owners were being harmed, so that, if harm *were* shown, a royalty could be established. Waz told a reporter that he deplored the "irrelevancies and charges and countercharges" being thrown about. "I really don't think the two sides are that far apart," he said, and he added that he was glad to be "able to play the role of the reasonable man." Six months later he joined Anne Wexler's PR firm, working on Betamax-related matters for the MPAA, and another copyright virgin had been deflowered.

By the middle of 1983 the two lobbying coalitions had assembled, between them, a vast Washington brain trust which included two former cabinet-level officials (J. Edward Day,* and Robert Strauss*), four former top advisers in the Carter White House (Anne Wexler,* Stuart Eizenstat,* David Rubenstein,* and Lloyd Cutler†); two former senators (Marlow Cook* and William Hathaway*); five former representatives (James Corman,* Peter Kostmayer,* Thomas Railsback,† Edward Patterson,† and Jerome Waldie†); two former chairmen of the Federal Communications Commission (Charles Ferris* and Dean Burch†); two former White House economists (Alan Greenspan† and Nina Cornell*); and two former chief counsels to the Senate Judiciary Committee (Emory Sneeden* and Ronald Brown*). Hardly any issue

* Retained by the Home Recorders or their constituents or allies.
† Retained by the Copyrightists or their constituents or allies.

facing the nation in recent memory had engaged the attention of a more formidable array of elder statesmanhood.

As the Betamax battle dragged on, it was possible to detect a change of attitude on Capitol Hill. The dignified duties of the lobbying profession and the valuable services performed in the past by the Home Recorders and the Coyprightists faded from people's memories. The word *Betamax* began to be associated with titters and whispers. Congressional aides told stories about the lobbyists behind their backs—comical stories of the sort one might tell about a wayward cousin or a mad aunt. Early in 1982 a lobbyist for the recording industry had telephoned a woman who worked for Patricia Schroeder, the congresswoman from Colorado, to ask if "your boss" would be willing to introduce the audio royalty legislation, intended, at the time, to go into a separate bill from the video royalty. The request was duly relayed to Schroeder, who duly agreed. But when the aide reported this news back to the lobbyist, he seemed flustered, and several days later he was on the phone again, explaining that the whole thing had been a mistake. When he had called Schroeder's aide, it turned out, he had meant to call (and thought he had reached) a woman who worked for Don Edwards of California, the sponsor of the video royalty measure; he had been after Edwards to introduce the audio bill as well. His mix-up led to considerable soul-searching among the recording industry's lobbyists. Afraid that Schroeder would be annoyed if they switched signals on her, they decided to go ahead with the arrangement as best they could. Edwards would introduce the video bill, and Schroeder (if still willing) would introduce the audio. Once again Schroeder agreed. But the solution came unglued when a higher council of lobbyists decided to combine the two proposals into a single piece of legislation. Even then an effort was made to keep Schroeder in the picture. Carlos Moorhead of California had been asked to serve as the prime cosponsor of the proposal, but the lobbyists went out of their way to attach Schroeder's name to it as well. Although two names per bill was the convention, this one would be known, by the lobbyists if by nobody else, as the Edwards-Moorhead-Schroeder bill.

"They spent forty days trying to figure out the name sequence on these bills—whose name goes second and whose name goes third," an aide to Schroeder said later. "That's the kind of thing lobbyists do."

From time to time they managed to cause annoyance as well as amusement. At a Senate hearing early in 1983 Senator Patrick Leahy of Vermont unloaded some pent-up feelings about the video dealers and the lobbyists who had been organizing them. "Let me give you a little bit of advice about that group, just as it relates to one small state, the State of Vermont," Leahy told Wayman. "I found their tactics to be deceptive, deceitful, outrageous, offensive, and that's giving them the benefit of the doubt. If it wasn't for the fact that we Vermonters tend to understate things, I would tell you what I *really* think about their tactics." One tactic that provoked Leahy was a "dunning Mailgram" purporting to come from a long list of Vermont video dealers who "demanded" to meet with him. When the signers of the mailgram were contacted by Leahy's staff, they "seemed rather surprised" and "weren't too sure what this Mailgram was that we were talking about," Leahy said. Just the same, he told Wayman, he had arranged a mass meeting with the dealers he contacted, and "of the thirty or so . . . two did show up, and fine people . . . I happened to know one of them quite well."

On another occasion a group of video dealers had come to a meeting with Leahy in Vermont accompanied by a public relations man from Washington, and the PR man, to Leahy's astonishment, had taken credit for helping "the people in my district" arrange the meeting. "Now, Vermont, just for a bit of history, was the fourteenth state to be admitted to the union," Leahy said to Wayman. "We tend to refer to it as a state, not a district. . . . Might I suggest to you that you will do yourself far more good if you will give those of us who wish to be responsive to our constituents an honest story. . . . I did not find the mailgrams that I got to be honest at all."

Senator Mathias, at the same hearing, sounded off against video dealers who were offering free movie rentals to customers who agreed to sign letters to members of Congress. "Are we going to have another of these battles of the computers," Mathias wondered, "with industry computers spewing out letters to the Congress and then congressional computers spewing them back?"

To that Jack Valenti replied that he found it "a bit ironic" that retailers were "using our products . . . to try to get people to write in to kick us in the pants."

It was the amount of money the lobbyists were making, and spending, that caused the most comment on the Hill. The Japanese electronics companies were the freest spenders by most estimates. Anxiety about the battle seemed to increase with distance from the scene. "Their side wound up hiring half of Washington," an aide to a proroyalty congressman said later. "It was one of the most lucrative things that ever floated through this hallowed city." According to Anne Wexler, the antiroyalty lobbyists were "scratching their heads because they couldn't believe how much money they were making." But Hollywood could not afford to operate on a significantly lower pay scale, and when detached observers fell to talking about the money lavished on this conflict, they rarely bothered to make much of a distinction between the two sides. "It's corporate pigs versus corporate pigs," was one congressional aide's assessment. "So you sort of watch it like a pig fight. All these people running around with five-hundred-dollar suits and three-dollar cigars. You walk into the hearing rooms and it's wall-to-wall pinstripes."

The lobbyists themselves were not unaware of the appearance they presented. "It's like an arms race—each side is fearful of the other side ramming through something fast, and so it keeps escalating," Charles Ferris told a reporter who had asked about the big names who had become involved. An anonymous lobbyist for the movie industry compared the two sides' use of experts such as Laurence Tribe and Nina Cornell to a murder trial. "The prosecution hires a psychiatrist who says the killer is sane," he said, "and you have to keep looking until you find someone who says your client is insane."

If there was an epicenter to the skepticism—a place where the Betamax affair was viewed more suspiciously than anywhere else—it was in the offices of Robert Kastenmeier, the chairman of the subcommittee through which any home taping legislation would have to pass before it came to a vote in the House. "Never go to sea," Sir Joseph Porter advised youngsters with the itch to become lords of the admiralty. The House's leading copyright authority scarcely gave the subject a thought until the early sixties, when, as a junior member of the Judiciary Committee, he was asked to oversee a series of hearings on revision of the 1909 Copyright Act. Before his election to Congress, Kastenmeier had been a lawyer and justice of the peace in a state known for cheese and beer, but he was a soft-spoken, intelligent, deliberate legislator who had won the trust of Emmanuel Celler, the chairman of Judiciary at the time, and copyright was such a marginal subject that Edwin Willis, Kastenmeier's predecessor as chairman of the subcommittee, was happy to yield the issue to a midwestern liberal while he focused on his own more stimulating duties as chairman of the House Un-American Activities Committee.

After a year of hearings a copyright bill was marked up in 1966 and passed the House in 1967. But the Senate wanted to chew it over for a little longer—for another nine years, to be exact—and while it did, the House was obliged to hold countless additional hearings and markup sessions in the constant hope of finishing the job. (The elderly daughter of John Philip Sousa—"in what I thought was a wholly unconvincing case," Kastenmeier recalled—was one of many witnesses who successfully pleaded that their ancestors' works not be cast into the public domain during the interim.) By the time a bill passed both houses in 1976, Kastenmeier was chairman of the subcommittee. By 1981, when the Betamax issue descended on him, he was a twelve-term veteran with graying hair and a few gentle wrinkles on his face, but with his patience intact.

Although not associated with any cause or bill that had gained national attention, Kastenmeier's name was linked to a way of conducting congressional business that had enjoyed some success by example. He operated in the open. He had been one of the first subcommittee chairmen to invite the press and public to watch a bill being "marked up." He listened to what witnesses said at hearings, asked questions that seemed to be intended to elicit information, and gave every indication of taking the answers into account. His own views were slow to develop—annoyingly so, to some tastes—and when they did, he rarely bothered to announce them.

Known or unknown, they were not often a topic of wide interest. His subcommittee's main concerns—courts and prisons—stood way down on the scale of congressional priorities, and its hearings rarely attracted the press. The Kastenmeier subcommittee certainly wasn't a stepping-stone to anywhere. Then again, Kastenmeier wasn't really going anywhere. If obscurity allowed him to run things his way—seriously, slowly, collegially—obscurity was all right with him. Just about the only problem he had with his subcommittee was filling it. "People didn't really seek to get on," he recalled, "and there were people who would get on and off."

Then, rather abruptly, a seat on the Kastenmeier subcommittee became a precious commodity. In 1981 and 1983 the demand so far exceeded the supply that Peter Rodino, who had succeeded Celler as chairman of Judiciary, decided to expand the subcommittee from seven to eleven and then from eleven to fourteen members. Kastenmeier viewed this wave of popularity much as the guardian of a homely heiress might view a porchful of handsome suitors. He wanted to believe the best but suspected the worst. The Judiciary Committee itself was by no means a plum assignment, what with the sensitive issues it faced—school prayer, busing, abortion, and capital punishment—and its "lack of pork," as one committee aide put it. The Betamax affair and a conflict over cable TV royalties (another issue pitting wealthy copyright owners against wealthy copyright users) not only were safer and more fun but involved corporations which could be counted on to reward a reasonably cooperative legislator with a campaign contribution. Kastenmeier could not help noticing that the surge of interest in his subcommittee coincided with its immersion in these controversies. Nor could he help noticing that two of his new members (Carlos Moorhead and Howard Berman) hailed from Southern California, while four more (Pat Schroeder, Romano Mazzoli of Kentucky, Barney Frank of Massachusetts, and Thomas Kindness of Ohio) wasted no time in speaking up for the movie industry and its fellow copyright proprietors. The possibility that Jack Valenti and the MPAA had had a hand in the expansion process crossed Kastenmeier's mind. Even if they hadn't, he suspected that there was a connection between the growth of his subcommittee and a growing awareness that, as he put it, "there's an awful lot of money in intellectual property."

"There's nothing more difficult for a member than raising money—

and nothing more unpleasant," Thomas Mooney, the subcommittee's minority counsel, explained. "So they really appreciate it when they get some help with that. This subcommittee was one of the better-kept secrets for many years—that there was gold in them hills. People wondered why anyone would want to serve on it. Now the reason is fairly well known. There's a whole community out there that is ready, willing, and anxious to contribute."

It was as if a disco had opened on the edge of Walden Pond. "For the first time people want to talk to you who aren't right-to-lifers," a subcommittee aide explained. "You can have a fund-raiser and just put out the word to Warner Brothers and MCA and so on. Plus you'll get invited to Jack Valenti's dormitory down there for movies and buffets, and if you want to see a movie, you just write down the name and they'll send it up to you."

Kastenmeier's subcommittee and the criminal justice subcommittee, chaired by John Conyers, had been traditional rivals for most barren territory, campaign contribution-wise, on Judiciary. Now Conyers's subcommittee had that honor all to itself, and one Judiciary staffer commented that Conyers's people were "beginning to feel like the Maytag repairman."

As time passed, Kastenmeier was pleased to note some "shifting of positions" among his new members and happy to learn that one whom he had suspected of joining "just to avail himself of campaign contributions" actually refused to accept money from either side in the home taping controversy. "I said, 'Great, I misjudged you,' " Kastenmeier recalled. "So I can't be too hasty. Some of them, I think, are really interested in the subject. They like making judgments about Hollywood, about the broadcast industry, about the cable industry. They like being in the center of a major contest." But sometimes, he added, their enthusiasm for such issues made it difficult to focus the subcommittee's attention on the unglamorous, and PACless, topics of courts and prisons, which also fell within its purview.

The member whom Kastenmeier misjudged was evidently Michael Synar, an Oklahoma Democrat. Synar had decided to refuse all PAC money from outside his state, and it proved to be a costly policy. He received $13,810 in PAC contributions for 1982, a year when most of his subcommittee colleagues hauled in at least five times as much, and when one of them, Barney Frank (whose contributors included the PACs associated with MCA, Warner Communications, Twentieth Century-Fox, and the Recording Industry Association) led the entire House of Representatives with $189,059.

Kastenmeier's own attitude toward PAC contributions did not escape reproach. He refused to take PAC money from industries with legislation pending before him, but as one Republican on the subcommittee wryly observed, lobbyists and company executives generally understood that any contributions they might wish to make as *individuals* would not be turned away.

Campaign contributions were one department in which the Copyrightists outdid the Home Recorders. But a subcommittee member could share in the bounty without making a definite commitment to either side. Just being there was enough. "It's a protection racket," one movie studio lobbyist explained. "Most of us don't think we get much of anything for our campaign contributions. It's just the price of doing business. When you become a lobbyist in this town, it makes you fair game. You get invitations to fund-raisers for members who have consistently opposed you, and you go, and the Sony guys are there, too. So you pay less—but you still pay something. You've got to pay to avoid being totally stomped on."

The Betamax issue had broadened horizons for the subcommittee's staff as well as its members. There are no revolving door laws for the legislative branch of the federal government, and it is de rigueur for congressmen to offer their profuse congratulations when staffers leave for more profitable situations in the private sector. When two of his top aides, Bruce Lehman and Timothy Boggs, left to become Hollywood lobbyists, Kastenmeier adhered to the tradition. He continued to invite them to his Christmas get-togethers, and he insisted, when asked, that their predeparture conduct had been "absolutely impeccable," although he quietly added that he hoped such job switches would be "more the exception than the rule."

During one of the early Senate hearings on the Betamax issue Robert Dole asked Jack Valenti if he had "tried to negotiate this problem with the Japanese."

"My door swings on easy hinges, senator," Valenti answered. "I think it is quite possible that if the six VCR manufacturers sat down with copyright owners a reasonable fee and a reasonable solution to this issue could be reached quickly." Senators Daniel Patrick Moynihan, Lloyd Bentsen, and Charles Mathias joined in urging Valenti to pursue the idea, and Bob Dole, in a jocular mood, proposed a possible formula. Why didn't the movie industry arrange with the videocassette manufacturers to enclose "a little notice" with each cassette advising the buyer, "If you use this more than once, please send in a dollar"? The money could be sent directly to Valenti or his designee, Dole suggested.

Valenti did his manful best to join in the wave of laughter that greeted Dole's proposal. Unfortunately the idea was only slightly more far-fetched than the possibility of quick Senate passage of a royalty. In the summer of 1982 the Judiciary Committee repeatedly scheduled markup sessions on the DeConcini bill and the Mathias amendment and never once got a quorum. Even senators who had proved friendly to the MPAA in the past, and who acknowledged the justice of its position in the present, showed very little lust for legislative battle. In all the arguments and counterarguments, the one thing that had impressed itself on all the members of the Judiciary Committee was the fact that *whichever* way they voted on this issue they would be making a lot of people angry. The Supreme Court's deliberations gave them the perfect excuse for not voting, and they intended to take full advantage.

The pace of deliberations in the House was even more frustrating to Valenti. Although a majority of the Kastenmeier subcommittee seemed to support the royalty, the chairman was being his usual slowgoing, noncomittal self. Kastenmeier didn't even *try* to hold a markup, and his colleagues, however ready to side with Hollywood when the issue came up for a vote, were not about to buck the boss on a procedural matter. So perhaps the time had come, Valenti decided in the summer of 1982, to give the idea of compromise a try. Maybe the industry could live with a smaller royalty than he had originally proposed. Maybe it could accept a royalty that covered only blank tapes and not VCRs—especially if the other side dropped its opposition on the rental issue. "We were really talking about a principle," Dale Snape explained. "To get the principle established, it was never contemplated that the motion picture industry would make much money out of this."

Valenti talked it over with Robert Strauss, and Strauss agreed to seek a meeting with Akio Morita. Strauss was a truce maker from way back. He had been instrumental in reuniting the Democratic party after the disaster of the 1972 election and in calming the Texas brouhaha that attended some unflattering remarks of Jimmy Carter's, during the 1976 campaign, about Lyndon Johnson. And he knew a thing or two about how to talk to the Japanese, from the Tokyo Round trade negotiations of 1979. As for Morita's receptivity, Valenti and Strauss knew he had not absolutely closed the door on the idea of a royalty when Harvey Schein, the former president of Sonam, had posed it as an answer to Universal's lawsuit. Of course, Morita could not speak for any company but his own; but he was something of a roving ambassador for Japanese industry, and he enjoyed a reputation in Japan as a shrewd judge of matters that concerned Americans and the unruffling of the feathers thereof. If Sony could be brought around, Valenti and Strauss reasoned that the other Japanese companies would probably fall into step.

Strauss sent out peace feelers through Robert McNair, the former governor of South Carolina, whose law partner, Emory Sneeden, had been retained by Sony. When the Home Recorders got wind of Strauss's designs, however, they counseled Morita not to meet with him. Hollywood's desire for compromise, they argued, was a sign of weakness and, thus, all the more reason to stand firm. Morita was reluctant to appear rude to "my good friend Bob Strauss," whom he knew from Strauss's days as U.S. trade representative, but he saw the wisdom of making himself scarce, and Strauss was forced to settle for a hurried conference with Kenji Tamiya and Sadami Wada of Sonam during a layover at La Guardia Airport. As McNair, the go-between, reconstructed the conversation, "We were trying to determine the parameters of the issue, and whether there were any areas we could come to a common understanding of—and it turned out there weren't. It was very cat-and-mouse because we all felt fairly comfortable with our positions and both sides were hopeful that they would win in the Supreme Court."

When Valenti and Strauss communicated the outcome back to their friends in the Senate, the bipartisan team of Dole and Bentsen decided to take a more direct role in the quest for compromise. Inspired by Japan's willingness to adopt "voluntary" quotas on automobile exports in the face of sterner restrictions threatened by Congress, the two senators dispatched a letter to Sousuke Uno, the head of Japan's Ministry of International Trade and Industry. They called for "face-to-face negotiations between the interested private parties" and explained:

> We bring this matter to your attention because it poses a potentially serious threat to U.S.-Japanese trade relations that you are in a position to avert. The U.S. trade deficit with Japan is projected to be nearly $30 billion this year. The size of this imbalance will exacerbate the perception widely held among Americans that Japanese imports compete unfairly in the U.S. market. Because the question of compensation for U.S. owners of intellectual property fundamentally is an issue of fairness, a Congressional decision on these copyright matters, taken in the context of a large Japanese trade surplus, is unlikely to be as agreeable to manufacturers of video and audio recorders and equipment as an agreement concluded on a mutually satisfactory basis.

The Dole-Bentsen latter alarmed the Home Recorders. Fearing an end run around their defenses, they urged the Japanese manufacturers not to give an inch and to let their government know that this was an internal American dispute, not a question of international trade. In Washington they alerted their chief Senate allies, DeConcini and Thurmond, who dispatched a joint letter of their own, assuring Minister Uno that "Senators Dole and Bentsen speak only as two members of Congress, and not for the Congress as a whole or the American Government. In fact, many members of Congress, including ourselves, would view any such action by the Japanese Government as pre-emptive of our prerogatives and hostile towards the American consumer."

A few weeks later Minister Uno informed all four senators that "it would not be appropriate for the Japanese Government, as a foreign government, to be involved" in a "domestic legal problem." And that was the end of that.

To Valenti the failure of these initiatives—the utter refusal of the opposition to consider them—was exasperating. The movie industry, it seemed to him, kept lowering its sights and bending over backward to be accommodating, while its adversaries only grew more belligerent. Valenti blamed the lobbyists as much as their clients. On several occasions Toshihiko Yamashita, the president of Matsushita Electric, had told the press that a royalty posed no particular problem for *his* company. If a royalty were imposed, he said, Matusushita would do its best to absorb the expense. Yamashita's views confirmed the predictions of the proroyalty lobbyists and their economic consultants. But they did not prevent Jack Wayman and his associates at the EIA from claiming at every opportunity that a royalty would be a devastating blow to consumers. Indeed, the public would have to pay *more* than the full cost of a royalty, according to Wayman, because of what he called "the doubling factor in distribution systems." A fifty-dollar royalty on VCRs, he said (always citing the highest figure that Valenti had ever mentioned), would translate into a hundred-dollar price increase because "just like on a car radio you have a markup at the manufacturer's level, at the distributor's-rep level and at the retail level."

Of all the positions taken by the Home Recorders, Dale Snape found this the most galling. "The copyright holder gets fifty dollars," he said. "Who gets the other fifty? What kind of business do those guys run? And they say they're on the side of the consumer!"

In the normal course of events nothing would have given Valenti more pleasure than the opportunity to work and speak for a cause in which he fervently believed. But the burden of responding to the Home Recorders' rhetoric was getting to him. Wayman and the EIA were especially grating. During one congressional hearing, Wayman referred to Stanley Gortikov of the Recording Industry Association as "Mr. Greedikov—er, Mr. Gortikov." Invariably he described the royalty as a "tax," explaining, in the event that anyone protested, "Well, I always say that if it waddles like a tax and it quacks like a tax, it's a tax."

"There's more misinformation and disinformation spread by that group over there than by the KGB, for God's sake!" Valenti complained.

"Debate on homevideo issues is reaching new heights of acrimony," *Variety* reported in August 1983. "In a town where even the most violent disputes are usually couched in careful terms, the latest phase in the homevideo battles seems to be spurring extraordinary levels of bad feelings, with MPAA prexy Jack Valenti dismissing the claims of EIA's Consumer Electronics Group senior veep Jack Wayman as the words of a 'stupid idiot,' and Wayman commenting that Valenti knows nothing about homevideo, and is "just mouthing what Hollywood tells him to say.' " *Variety* concluded that "the mud is being slung so deep some fear Congress may fear to tread in it. . . ."

Valenti finally resolved not to appear with Wayman anymore. "I didn't mind being with Ferris," he said. "He was the spokesman. But I wouldn't get up there with some public relations fellow. Rather than debate with Wayman, I would send my public relations guy to debate with him." Even without Wayman to contend with, the task of defending the movie industry's position was a challenging one. His debates with Ferris often ended with the audience taking a vote, and, according to Valenti, "Charlie learned—too quickly—that if he called for a vote, he was bound to win because his argument was simple to state. Charlie would say, 'Nobody ought to be taxed for taping in their home.' That's one sentence. Now, to refute that, it takes me fifteen minutes to go into the whole tormenting journey of how you invest in a movie, how you rely on posttheatrical markets, the history of copyright and why it's important. I tried to compress it all down to one page and commit it to my memory. It was impossible to do. I was like a racehorse carrying fifty pounds in lead pellets in my saddlebags, and Charlie was carrying none."

The spoken word was Valenti's specialty. He had just published a book on the subject *(Speak Up with Confidence: How to Prepare, Learn, and Deliver Effective Speeches)* which carried endorsements from Edward Bennett Williams and William Safire. Yet in the home taping conflict it was the other side that had succeeded in defining the terms of debate. Again and again the advocates of a royalty stressed that they spoke for documentary filmmakers, writers, unemployed actors, and others who were by no means sitting pretty. But nothing they said or did could shake their cause free of the label "Hollywood" and all it implied: limousines, hotel suites, cocaine, and the rest.

"For the studios it's Christmas in May," Wayman commented during one of his joint appearances with Valenti. "They're having an all-time record year in profits, so somehow this vulture, VCR, is certainly not ruining Hollywood's revenues at the box office."

But when Valenti tried to call attention to the immense profits of the Japanese electronics companies, his words had a way of bouncing back at him. He regarded the royalty as a necessary defense of American creativity against Japanese exploitation. "We're talking about a massive trade asset to this country," he would point out. "American audiovisual material—movies and television—return to this country almost a billion dollars a year in surplus balance of trade. We're one of the few products in America that has a trade surplus with Japan, for God's sake! I guarantee you that if it were reversed, if the Japanese dominated the world in television and movies and the Americans were exporting machines to Japan, gobbling up and ruining the value of Japanese-created material, MITI and the Japanese Diet would be passing laws to protect *their* creative people."

Once started on this theme, however, Valenti tended to drift into phrases that were susceptible of misunderstanding. "The American movie is the one thing the Japanese with all their skills cannot duplicate or clone," he would say. "They can't make pictures like we make them." At one congressional hearing Valenti defined the issue as "the American people versus six Japanese manufacturers." At another he referred to Ferris as "one of the Japanese lobbyists" and looked very surprised when a portion of the audience began to boo him. "Well," said Valenti, "if I am saying something wrong, please forgive me. . . ." On still another occasion he tried to turn the tables by complaining about the anti-Hollywood "racism" of the Home Recorders. He was tired of hearing them talk about Hollywood "as if it were something dirty and squalid." After all, he said, "Japanese machines do not create entertainment. The American motion picture industry does." Emotions cooled when Ferris replied that Hollywood was "a place of great affection" for him, and added, "I'm sorry that you think it's racist when I talk of Hollywood, Jack."

DECISION

ON THE FIRST Monday in October 1983, the Supreme Court held its second hearing in the matter of *Universal* v. *Sony*. Dean Dunlavey and his wife and secretary flew into Washington on Saturday, in time to have dinner with Ira Gomberg, Teruo Masaki, and Bob Bruce, the three lawyers in Sony's employ who knew more about the case than anyone other than Dunlavey himself. They ate at 209½, an elegant little restaurant on Capitol Hill. As the only Washingtonian of the party Bruce had chosen the restaurant, and he took note of Dunlavey's order: steak, orange juice, and ice cream. On Sunday the four men got together again for a briefing session. It was Dunlavey who did most of the briefing. "Dean is not a consultative person," Bruce later explained. "He had a very clearly focused view of what he wanted to do."

Going into the reargument, Dunlavey assumed that the justices were "familiar with time-shifting and the fact that you can copy one program while viewing another and that their thinking was now at some elevated level where they saw complexities they did not see the first time around and would want to inquire into them." He expected questions to rain down on him like artillery fire. Instead, the hearing began with Chief Justice Warren Burger politely asking him to "proceed whenever you're ready," as if it were Dunlavey who had requested another appointment with the Court and not the other way around.

He began by treating the justices to a short course in the history of American television. "The small screen has gone to the large screen," he said. "Black and white has gone to color, tubes have gone to transistors, and live broadcasting has gone to delayed broadcasting, using pretty much the same VTR that we're concerned about here this morning. The Court can remember that the early television shows all had to be broadcast live because there was no means of recording them." Sony, he told the justices, had received an Emmy Award in 1976 for its U-matic videocassette recorder, which had become a basic tool of TV news. "And now progress has continued," Dunlavey went on. "That same VTR has been put in the hands of the public so that, instead of a necessary live viewing, there can be a delayed viewing. And the question this morning is, Can the copyright owners or a few of them who are profiting by exploiting their product on free television arrest this latest progress of science?"

Normally the facts at issue in an appeal are those set forth in the trial judge's opinion, and lawyers and judges alike must pretend that history stopped there. But a lawyer is permitted to introduce "common knowledge" into his argument, and a judge may, in his conclusions, take "judicial notice" of certain inescapable truths. In the reargument Dunlavey made liberal use of these concepts in order to acquaint the Court with the dramatic growth of the video business in the preceding few years. If the law was ambiguous, the justices would have to look outside the law to reach their decision, and if they did, he wanted to be their tour guide, showing them the Betamax as an ingenious, versatile, popular invention in the great American tradition of discovery and showing them the executives of Universal and Disney, the plaintiff companies, as men trying to defend the old against the new.

"I don't think anybody has trouble with the concept that you really shouldn't be using the copyright owner's works against him, all other things being equal," Dunlavey said. But he asked the Court, "Let's see what is the effect on him? Now the Ninth Circuit with no explanation simply says, 'It seems clear that home recording does tend to diminish the potential market for the Universal-Disney works.' No explanation. Just *ipse dixit*. . . . The fact of the matter is, and it's common knowledge and judicial notice today, that the prerecorded movie is available across the country. . . ." Watching prerecorded cassettes, he said, constituted "thirty percent of the primary reason why people are using video tape recorders—thirty percent." He didn't bother to say that prerecorded cassettes could be played equally well on a machine without a recording capability. "They complained that there would be a decrease in the future audiences of their television product and their motion picture product. In fact, Mr. Sheinberg . . . took the stand and in all seriousness said this will be the ruination of the movie industry." But "MCA has just reported its highest six months' revenue in history. . . . There is no sense in wandering off into a hypothetical assumption of horribles as to what might be done with the tapes. The fact of the matter is that they're staying within the household, only people within the intended audience are seeing them, and then they're being erased, and absolutely no harm is being done. . . . The prerecorded cassette has been a bonanza in the hand of the studios. The net effect on the studios . . . has been a substantial benefit, certainly no detriment."

Stephen Kroft—"his approach . . . sincere, his manner calm, his presentation thoughtful," according to a critique in the *American Lawyer*—laid out his case in equally broad strokes. "Underneath all the legal arguments and legal labels that we've thrown around in this case, the case is really very simple and straightforward," he began. "Petitioners have created a billion dollar industry based entirely on the taking of somebody else's property—in this case copyrighted motion pictures, each of which represents a huge investment by the copyright owners. And contrary to what petitioners would have the court believe, respondents are not the only copyright owners who have raised express objections to these activities. The Court has before it *amicius* briefs raising such objections from over seventy copyright owners, including such people as the CBS television network, the producers of approximately ninety percent of the programs on prime time network viewing, and many, many producers of educational and cultural programs, such as the Children's Television Workshop, the Lincoln Center for the Performing Arts, and the producers of the National Geographic series and the Smithsonian Institution specials."

When the justices weighed in with their questions, however, Kroft answered with firm interpretations of law. What would he say, Justice White wanted to know, if fifty percent of the programs on television were uncopyrighted or owned by producers who had no objection to home taping? "I'll go you one better, Justice White," Kroft replied. "If there was only one show on the air that were copyrighted and which could not be copied without objection, if the petitioners sold this device with knowledge that it would be used to copy that show, under the Inwood test laid down by this Court in the trademark area, I believe the petitioners would be liable."

Justice John Paul Stevens asked Kroft if he would make the same argument for photocopying. Was it his "view of the law that as long as Xerox knows that there's some illegal copying going on, Xerox is a contributory infringer?"

"To be consistent, your honor, I'd have to say yes," said Kroft.

"A rather extreme position," Stevens commented.

At the time, Kroft seemed perplexed by some of the questioning, but he looked back on his two appearances before the Court as "absolutely exhilarating" because "every member of that court fully and completely understood the issues in the case and genuinely wanted to illuminate the issues as best as they could without trying to grind an ax."

Dunlavey maintained the same low opinion of the Court he had formed after the first argument. The justices should have been "up there like sharks with a bunch of hard, pointed questions," he explained. "At neither of our hearings did we get those questions."

Early in the reargument Justice Sandra Day O'Connor asked if the Court could resolve the case on the contributory infringement issue alone—by deciding that Sony wasn't responsible for what customers did with their Betamaxes—without determining if home recording itself constituted infringement.

"Justice O'Connor, that's precisely right—there are two roads to Rome," Dunlavey replied. But as he answered, he was thinking: "For Christ's sake, I've been telling you that since the day I first knocked on your door with my petition for certiorari. Why has it taken you until my second argument to perceive something that I broadcast across my brief when I first filed it?"

Toward the end of 1983 the lobbyists took up their vigil at the Court again. "We have felt like the anxious father waiting outside the delivery room," said J. Edward Day, "but now it's like the delivery room at the zoo because only an elephant has a gestation period of fifteen months."

James Corman, the former California congressman who had become a lobbyist for MCA, had a dream about the case one night. He dreamed that the Supreme Court had handed down a decision affirming the lower court. In the dream Corman was ecstatic. But on waking he realized he had forgotten to ask, "Which lower court?"

The lobbyists speculated endlessly about the result, and it was speculation almost completely untainted by hard facts. From the justices' questions and comments Kroft believed that two of them had been leaning toward Sony and one toward Universal. "But you know, litigation is a funny thing," he said. "Sometimes you read the wrong signals. One signal to read is that if they took the case to reverse it, they could have done it a long time ago." He seemed weary of the lawsuit. "I was thirty-two years old when I filed it," he said. "I'll be forty years old next week. What is that? Twenty percent of my life? And if we win, that won't be the end of it."

At a few minutes after ten on the morning of January 17, 1984—three and a half months after the reargument; seven years after the case had been filed—word filtered out of the public information office at the Supreme Court that one of the decisions to be handed out that day was *Universal* v. *Sony.* Toni House and her staff were sitting at their desks, with a well-burnished wooden railing guarding them from the fifty lobbyists or so who had swarmed into the office. The distribution process was simple: One of the Court employees would raise a hand with a copy of the decision in it, and the copy would be plucked away. Once in possession of their treasure, the lobbyists showed all the patience and decorum of Sidney Greenstreet, Peter Lorre, and Mary Astor on receipt of the purported Maltese falcon. As Michael Blevins, who was there on behalf of the EIA, recalled, "A lot of sedate, well-dressed people got very pushy and grabby." The lobbyists had been prepared for the possibility that some study and expertise might be needed to know just what they had. Instead, a glance at the first page was enough.

"We now reverse," Justice Stevens wrote, speaking for a bare majority of five justices, the others being Brennan, White, Burger, and O'Connor. "One may search the Copyright Act in vain for any sign that the elected representatives of the millions of people who watch television every day have made it unlawful to copy a program for later viewing at home, or have enacted a flat prohibition against the sale of machines that make such copying possible."

Kenji Tamiya, the president of Sonam, received the news in New York at twenty minutes after ten and promptly called Akio Morita in Tokyo, where it was twenty minutes after midnight. "Of course, he is asleep, but I wake him," Tamiya said later. "When I told him about the decision, he asked me, 'Is that right?' I told him, 'I congratulate you.' And he congratulated us—Sony, U.S." Tamiya said he hoped Morita didn't mind being awakened. "Absolutely not," Morita replied. He would be awake all night. "Mr. Morita doesn't drink," Tamiya said later. "He drank two cans of beer—and still he couldn't go back to sleep."

Breaking from the old practice of using American executives as spokespersons, Tamiya read Sonam's official statement at its press conference, although Ira Gomberg stood at his side and fielded most of the questions. "This has been a long and difficult battle for Sony," Tamiya said, "but we are extremely gratified by the outcome." Because "there may be efforts by some to undermine this decision in Congress and through further litigation," he added, ". . . consumers must inform their elected representatives of their views on this important issue."

By eleven that morning most of the Copyrightists' regular breakfast group was sitting around a table at the MPAA discussing the decision and, in Dale Snape's words, "how to project it." It was not an easy assignment. "We missed 'Entertainment Tonight,' " Snape said, "because they have a twelve-thirty deadline—I don't know why, but they do—and they wanted Valenti, and we weren't ready."

It was midafternoon before they *were* ready. Valenti told a roomful of reporters that the five to four vote was "ample evidence that the court is troubled and divided." Even the majority, he observed, had invited Congress to consider the issue. ("It may well be that Congress will take a fresh look at this new technology," Justice Stevens had written.) This made the decision "an opening shot" in a continuing battle. It was up to Congress, Valenti said, to "decide whether . . . copyright is real, or whether it is mush."

Two days later Dean Dunlavey was accepting congratulations uneasily. "That was too close!" he said, wiping his brow. "Do you realize there were four judges who would have made Sony pay statutory damages for every infringement? I feel like I've walked a tightrope across the Grand Canyon, and now that I'm over I see where I've been."

Although the Court had divided narrowly, the majority considered the case clear-cut. The Betamax was the "staple article of commerce" that Dunlavey had claimed it to be. Universal and Disney were only two among many owners of copyrighted TV programs, the majority reasoned, and it would be wrong to let their views govern the conduct of so many others. The opinion attached particular importance to the fact that Fred Rogers, the star and creator of "Mister Rogers' Neighborhood," had conferred his blessings on the Betamax. Rogers's program, the Court noted, "is carried by more public television stations than any other program" and reaches "over 3,000,000 families a day." So long as some copyright owners felt as Rogers did, it seemed to the Court that "the business of supplying the equipment that makes such copying feasible should not be stifled simply because the equipment is used by some individuals to make unauthorized reproductions of respondents' works." The plaintiffs' arguments had made no allowance for an artist who cared more about the size of his audience than the amount of his earnings. "Both Hollywood and high-tech industries spent millions of dollars on high-powered lawyers and reams of briefs," an anonymous commentator wrote in *Copyright Notices*, the in-house journal of the federal Copyright Office. "Yet the testimony of Fred Rogers . . . may have tipped the scales."

In retrospect, it seemed that Sony had benefited from some extremely fortunate timing. Rogers and the other nonobjectors had been rounded up when Universal, to all appearances, was out to forbid home taping rather than set up a system of compensation for it. If the nonobjectors had been aware of the royalty option, some of them might well have found it attractive. Indeed, a few weeks after the decision two of the nonobjectors—the National Basketball Association and the Faith Center Church of Glendale, California—formally renounced the depositions they had given Sony. Now, when it no longer mattered, they objected to home taping after all. Fred Rogers's view, however, remained unaltered. "I've always thought of the 'Neighborhood' as a chance to have a ministry with families," he explained, "and I'm just very grateful to be allowed to do it. I've been given a great gift. There are many compensations for what I do."

As Justice O'Connor had suggested during the reargument, the Court might have decided the case on the staple-article-of-commerce question alone. It did not have to rule on the legality of the conduct for which it had absolved Sony of responsibility. But the majority was in an expansive mood. Justice Stevens's opinion volunteered that the individual home taper was also innocent, even if he did nothing with his Betamax but copy Universal and Disney productions. Time-shifting, according to the opinion, was a "noncommercial, nonprofit" activity with "no demonstrable effect on the potential market for, or the value of, the copyrighted work."

Among the legal authorities who assessed the opinion for the benefit of the press, there was a feeling that it was bad news for the proprietors of other forms of intellectual property threatened by new technology. Computer software makers, it was thought, would now have a hard time prosecuting anyone who sold a program designed to help consumers defeat copy prevention techniques. Dunlavey was perfectly willing to acknowledge the legitimacy of these concerns. "But my interest is in winning the case that I'm given," he added. "As to how you go about protecting all the different kinds of copyrighted material, that's for the academicians."

He couldn't really talk about the implications of the opinion, however, because he hadn't read it. Scores of copies had gone out to lobbyists and reporters in Washington. Gary Shapiro had telefaxed four hundred to the membership of the EIA's consumer electronics group. But no one had bothered to get the text to Dunlavey, and that was all right with him. *Universal* v. *Sony* was yesterday's business. He had already thrown away his Betamax scrapbook, and when asked if it had been the most interesting lawsuit of his career, he responded, "Not as interesting as the next one."

Supreme Court majorities don't come any smaller than five to four, and when one is on the short end of that margin, it is hard to resist thoughts of what might have been. It was especially hard in *Universal* v. *Sony.* Justice Harry Blackmun's unusually long dissent—forty-four pages—opened with an elaborate recitation of the facts of the case. Supreme Court experts were quick to note that it "bore all the earmarks of having been drafted originally as a majority opinion," as Jim Mann observed in the *Los Angeles Times.* Blackmun himself bore all the earmarks of an unhappy jurist. Twelve years earlier he had been obliged to abstain from voting in *Williams and Wilkins,* the landmark copyright case which involved the photocopying of articles from medical journals. Reduced to eight participating justices, the Court had split, four to four, thus affirming the lower court ruling against the copyright owner. The Betamax case was an opportunity for the Court to correct what many copyright experts regarded as a grievous error, and Blackmun apparently shared that view. At the outset of the Court's business on the day the Betamax decision was announced, Justice Stevens delivered an oral summary of the majority opinion, and Blackmun responded with an explanation of his dissent. Stevens, Blackmun said, had made "an eloquent and almost emotional plea on behalf of the Court's judgment," but his intensity struck Blackmun as an attempt to obscure an unwillingness to enforce the law against popular opposition. The majority spoke of the Court's "consistent deference to Congress" in the face of "major technological innovations." Blackmun thought it would be more accurate to say that "the Court has tended to evade the hard issues when they arise in the area of copyright law. I see no reason," he went on, "for the Court to be particularly pleased with this tradition or to continue it."

The justices had divided along unusual fault lines. The liberal and like-minded duo of William Brennan and Thurgood Marshall had gone opposite ways—Brennan with the majority, Marshall with the minority—and the four most conservative justices had split, with Sandra Day O'Connor and Warren Burger taking Sony's side and William Rehnquist and Lewis Powell Universal's. The fact that Blackmun had parted company with his old friend Burger was no longer the novelty it once would have been. Burger had been the best man at Blackmun's wedding, and in their early years on the Court they had been known as the "Minnesota twins." Lately Blackmun (as he revealed in a series of not-quite-on-the-record speeches to select audiences) had come to regard Burger as a political operator unwilling to take risks to defend the integrity of the law, and the chief justice, for his part, had almost stopped assigning significant opinions to Blackmun.

Despite all these troubles, the circumstantial evidence suggested that Blackmun had expected to speak for the majority, not the minority, in the Betamax case. Someone on whose vote he had been counting had been struck by doubt in the final days of the 1982–83 session, it seemed, and had wound up voting the other way six months later. In the plaintiffs' camp Justice White was viewed as a likely defector because he had seemed so sympathetic to Universal's position during the first argument. On equally insubstantial evidence, however, a case could be made—and was—for Justice O'Connor, who had been drifting on the same ideological currents as Burger in recent years.

Dunlavey had nothing to say about any of this. But the disagreement between the majority and the minority struck him as curiously neat. Blackman and Co. had found the home taper and his supplier equally culpable; Stevens and Co. had found them equally innocent. "I'll bet you a dollar to a doughnut," said Dunlavey, "that there was a time during this year and some months that they've been sitting on that case where several, if not more, of those justices had a divided opinion—that there was direct infringement but not contributory. There just have to have been some people that crossed that clean-cut line. I figure there must have been so many multiple concurrences and partial dissents and what have you that when they looked at them, they realized they wouldn't be giving any guidance to the lawyers and the lower courts at all. They realized that the whole country was looking down their throats, and they figured they had to give the people a decision that would provide some guidance. On *Williams and Wilkens* they had split, four to four, and it was brought home to them time and time again that they had left things in a quandary. So I figure there must have come a time when they said, 'Somebody will write a white opinion and somebody will write a black opinion, and you can pick one or the other.' That would have been what [Earl] Warren would have done if he was the chief judge, and I have a feeling that something akin to that happened this time. As to how many were on one side as against the other side at any particular time, nobody but the Court knows, but thank God that when they reached the wire"—he laughed—"it was five on Sony's side."

Not long after the Betamax decision came down, a group of younger lawyers at Gibson, Dunn, and Crutcher—Dunlavey's firm—got into a bull session about whom they would want representing *them* if they were in *deep* trouble. The debate came to a close when someone said, "Knock it off, you guys. You all know you'd go to Dean Dunlavey." Even Sidney Sheinberg, in his way, was impressed. "I would never have felt comfortable with a Dunlavey type of lawyer," he said. "He's a blustery bullshitter, frankly. But he won."

Universal v. *Sony* had left a definite imprint on Sheinberg's image in Hollywood. Although he remained a powerful and respected figure, his belligerent attitude toward the VCR had become increasingly peculiar—simply by staying the same. Most of the studios had been through changes of management in those years, and the incoming executives tended to take a fairly unemotional view of home taping. To them it was just another part of the economic landscape. Even Disney had distanced itself from its coplaintiff by refusing to take part in a second lawsuit in which Universal lodged essentially the same charges against forty-two new defendants—eight foreign manufacturers of VCRs, eighteen American companies involved in marketing them, and sixteen advertising agencies. Disney had "come to realize that the interest of all concerned can be better accommodated by the passage of new laws," the company had stated in a press release intended to explain its nonparticipation. (Jack Wayman had come up with another explanation. "Mickey Mouse is getting a black eye," he had said.) Disney, along with Fox and Warner, had been one of the early studios to enter the home video market. It had struck a deal with Sony, of all companies, to distribute its cassettes, the relevant Disney executive telling the relevant Sony executive (who had raised the issue), "Look, the people who are suing you are a different division of the company. If this makes economic sense for both of us, why should we let the lawsuit stand in our way?"

In October 1982 Sheinberg had withdrawn as cochairman of a benefit for the American Film Institute's new video center upon learning, to his "shock" and "painful disappointment," that the center would get three hundred thousand dollars and a generous supply of video equipment from Sony. "I cannot condone nor support any institution which is so oblivious to the interests of the Hollywood community," Sheinberg had stated in his letter of resignation. Elsewhere in the Hollywood community, however, the news of the AFI-Sony relationship had been received with remarkable calm, and some people had shown more interest in Sheinberg's response than in the circumstances behind it. To file a lawsuit was one thing; the studios were constantly filing lawsuits. But to let a legal dispute intrude on a social occasion—*that* struck some of Sheinberg's colleagues as very odd.

The outcome of the Betamax case was, of course, a blow to Universal's reputation as a tough customer in court, and it was not the only blow the studio sustained in that period. In December 1983 a federal judge in New York had summarily dismissed a lawsuit filed by Universal against another Japanese company, Nintendo, which had developed a video game called Donkey Kong. The title character was a giant gorilla—*donkey* being a transliteration of the Japanese word for *goofy*—who held a woman captive atop a building while the player attempted to rescue her. Universal argued that Donkey Kong infringed on the King Kong trademark, which it claimed to have acquired in the settlement of an earlier court dispute with RKO and Dino De Laurentiis. But the judge, Robert Sweet, concluded that Universal had no legitimate rights to infringe, and the case did not rest there. Nintendo had filed a countersuit for more than a million dollars in legal expenses, plus untold millions in lost revenues from contracts (for dolls, hats, shirts, breakfast foods, and the like) supposedly frustrated by Universal's warnings to potential sublicensees. In the course of discovery proceedings in the countersuit, Judge Sweet learned that two of Universal's own lawyers had challenged the validity of its King Kong claims. By suppressing that information and telling Nintendo that it was the "sole and exclusive owner" of merchandising rights connected with King Kong, Universal had engaged in deliberate deception, in Judge Sweet's view. "Exemplary damages," he wrote, "are therefore justified."

Judge Sweet had equally harsh words for Universal's treatment of Coleco, an American toy company, best known for the Cabbage Patch doll, which had obtained a license from Nintendo to sell a home version of Donkey Kong. Sheinberg had persuaded Coleco to give Universal a three percent commission—ultimately amounting to almost five million dollars—for the rights the court ruled to be nonexistent. After the revelations in the Nintendo trial, Coleco sued for the return of its money, and Judge Sweet's decision seemed to bode well for its prospects. The legal basis for Universal's conduct in the Donkey Kong affair struck Judge Sweet as so threadbare that the case could be understood, he believed, only through an understanding of Sidney Sheinberg. He was the "Willie Keith of this story," Judge Sweet wrote in his opinion. " 'The event turned on his personality as the massive door of a vault turns on a small jewel bearing.' (Caine Mutiny, p.1)."

During his testimony in the Donkey Kong trial Sheinberg alluded to a joke that was making the Hollywood rounds: The two things inscribed on his tombstone, it was said, would be "King Kong" and "Sony." The joke did not seem to have caused him any great anguish. "Maybe I've become known as an excessive litigator, personally," Sheinberg said a few months later. "There are always people around who say, 'Why fight for something you can't win that much on? Why don't you devote your energies somewhere else?' If I'm the president of this company, I'm the trustee of the assets of our shareholders. I think I not only have a duty to try to maximize the revenues of today's endeavors, but a duty to protect the library which are the assets of this company. And I don't know how else to do it other than, in some cases, through litigation."

RETREAT

TWO AND A half weeks after the Supreme Court had decided the Betamax case, a group of senators, representatives, and congressional aides checked into the Bonaventure Hotel in Fort Lauderdale, Florida, for a three-day symposium on the problems of copyright and new technology. The symposium was the joint creation of Charles Mathias, who had finally been named chairman of a reconsecrated patents, copyright, and trademarks subcommittee in the Senate, and Robert Kastenmeier. Mathias and Kastenmeier did not normally see eye to eye, but they agreed that it was difficult for Congress to deliberate wisely in a field so bewilderingly complicated and so foreign to the experience of the typical middle-aged legislator. The two chairmen hoped to become better acquainted with the new gadgetry they kept hearing about, and to ponder its legislative implications in a setting free of lobbyists, journalists, and the other bric-a-brac of daily life in the nation's capital.

A panel of experts had been assembled for the occasion, and they told horror stories about the "stylishness" of copyright piracy. College professors, according to Alexander Hoffman, an executive of Doubleday and Company, "would go into a copy shop and say, 'I've got sixty students in Sociology 12. I want you to make sixty copies of this material—two chapters from this book, three chapters from this book, one chapter from this book. . . . I want you to bind them. I want sixty sets, and I'm going to sell them to the students, say, for thirty-five or forty dollars.' All of this without permission, all of this without payment. Over the years," he added, "we've published lots of anthologies. We can't do that very much any more, because they are all stolen."

Another panelist reported that business executives charged with buying computer software for their companies were saying to themselves, "Look, let's buy one set. We have a hundred and fifty people who need it. We'll make a hundred and fifty copies."

In the computer video game field, piracy was said to be so rampant that there were thirty unauthorized copies of a piece of software for every authorized one. Perhaps a third of the cable TV viewers in New York City were said to have tapped into the system without paying for it. Some of these people, according to Mel Harris of Paramount Pictures, were telling their friends, "You mean you're paying? Well, fool, why are you doing *that?*" Computer software companies had tried to defend against the pirates with encryption codes, but these were almost invariably overcome by disencryption codes, such as one wittily named CopyWrite, which came with a guarantee: If a piece of software proved uncopyable, the customer could mail it in, along with his CopyWrite diskette, and get back an altered version of CopyWrite that would do the job properly. (The manufacturer claimed that CopyWrite was intended merely to help people make backup copies of software for their personal use; thus, in the event of a lawsuit, the company was ready to claim that its product, like the Betamax, was a staple article of commerce.) In the computer world, as in the video world, a race was on between the encryptors and the disencryptors, and one participant in the symposium unhappily asked "whether we want to encourage the development of forces that will devote fine minds, and much valuable time, to the production, and destruction, of ever more elaborate encryption safeguards—minds that might more productively be devoted to the development of new, positive programs instead."

The outcome of the Betamax case had spread a canopy of gloom over the copyright community. It was seen as a threat to on-line computer services, cable TV companies, and many other commercial concerns which hoped to transmit information to customers in a convenient form and yet prevent the phenomenon of downloading—the retention of a permanent copy by a supposedly temporary user. Once a lawful copy existed, the first sale doctrine placed extremely narrow limits on the copyright proprietor's ability to restrict its use. In the opinion of some copyright people, the Betamax decision and allied developments had undermined a basic concept in the financing of movies, computer software, and other expensive creative works—the notion of collecting a small sum of money, tollbooth-style, at regular intervals along the road.

Harlan Cleveland, the director of the Hubert H. Humphrey Institute for Public Affairs at the University of Minnesota, suggested that something even more fundamental was at issue: "the notion that facts and ideas are owned." It might be necessary to look for nonmonetary ways of rewarding creativity, he suggested; as an example, he noted that the academic world uses "promotion, tenure, and tolerant attitudes toward minimal teaching loads and outside consulting." Another speaker pointed out that painters don't receive royalties every time their works are sold or displayed and seem to keep on painting just the same. (Perhaps, he might have argued, that's why they keep on.) Other panelists suggested that what technology had put asunder, it could put back together. Gustave Hauser, one of the founding fathers of cable TV, argued that the information age was at "an interim step in technology, one that still permits thievery and copying," and in the process of "evolving to something else, if we can only wait for it." The something else was "addressability," which he defined as "a capability to address specific programs, whether audio or visual, to specific users and to secure it at the subscriber's premises."

But Hauser's prediction provoked fears about the privacy and First Amendment implications of such systems and observations that the American people—who had shown a marked preference for the VCR over cable TV and for the personal computer over the mainframe—did not always accept the form of technology the experts decreed, no matter how superior it might appear from a copyright protection standpoint.

One man's piracy can be another's "cultural value of sharing," Frederick Weingarten, a program manager with the Office of Technology Assessment, observed. In the early days of computers, Weingarten recalled, hardware manufacturers had provided little in the way of software, and users, forced to fend for themselves, had passed around what they had, refining each other's refinements and becoming, along the way, software designers themselves.

Stephen Breyer, a copyright scholar who had been elevated to the federal bench, warned against "the evils of restricting copying . . . of requiring permissions, going through a lot of red tape to get permission to copy," and paid tribute to the "enormous social benefits obtained by copying." As a teacher he added, "Everything I say, I want copied. It should be copied, if anybody'll copy it. The problem is, they won't."

Throughout the discussion Robert Kastenmeier remained on input mode. A couple of days of listening didn't faze him, and if the talk tended toward the abstract, that was also no problem. The 1976 Copyright act, he realized, had failed to resolve a host of problems, and he foresaw a series of replays of the Betamax battle as other new devices aroused the anxiety of other intellectual-property owners. He was on the lookout for a better way to approach these crises than simply taking up the pleas of each aggrieved party in turn. One of his favorite questions, of late, was whether Congress should craft laws that try to anticipate technological change, or be content to react after the fact. He had been impressed with the arguments of David Lange, a copyright maverick at Duke University, who worried that the "loudest voices" in copyright disputes "tend to be raised in the service of particular copyright interests rather than on behalf of the public domain." Comparing some copyright proprietors to "greedy children," Lange had suggested that "sometimes greedy children have to be dealt with by saying no." He had proposed "a kind of civil procedure for new copyright legislation—a system imposing the legislative equivalent of burdens of proof and adverse presumptions to be met by anyone who proposes to extend the scope of existing copyright protection or who proposes protection for a new interest."

Congress, it seemed to Kastenmeier, had got into the habit of letting commercial interests set its agenda. "The law is no longer purely a matter of rewarding authors or creators," he said during an impromptu interview between sessions. "It becomes corporate interests fighting. For example, what is there about a copyrighted National Football League game that suggests a creator or an author? Really, nothing. I'm not saying it shouldn't be protected, but copyright law may not be the best way to protect it. We need to step back and see where we are—where technology is taking us. One of the concerns I've had is, even if the great industries agree, is the public interest also served?"

Charles Mathias's thoughts were running in quite another direction. Over lunch on the second day of the symposium Mathias talked about a growing split between "captive consumers," who pay for information, and pirates, who don't. He said he was afraid that the costs of piracy were being "uploaded" onto the captive consumer. Wasn't there a Catch-22 in the call for slow and careful deliberation on these questions? "Congress is already tender of the feelings of the ten million people who have VCRs today," Mathias noted. "Think of what it'll be like when there are eighty million. Isn't there some point at which the problem becomes so massive there's really nothing you can do about it?"

Congress had *tried* to anticipate change in the 1976 act, Mathias insisted. The dictum against unauthorized copying, as he understood it, was supposed to cover future technologies as well as present ones. As one who had helped craft the law, he could not see how the Supreme Court could have read its intent any other way. "The last previous copyright revision was in 1909," he said. "We thought we were legislating for another couple of generations at least. We even disbanded the copyright subcommittee when we completed that law. On the seventh day we rested."

Sitting at a lunch table crowded with Senate aides and other relative youngsters, Mathias harked back to the sixties, when he had served on the House Judiciary Committee under Emanual Celler. In those days, he said, the only revenue a songwriter received for the playing of his work on a jukebox was his royalty on the sale price of one 45 rpm record. At ASCAP's request Celler took on the mission of fighting for a compulsory license and royalty fee for jukeboxes, and it was "an interesting political science case," Mathias said, "because he would get a bill approved by the House Judiciary Committee, and then my recollection is it would go to the House Rules Committee, which was dominated by southern barons, and there it would disappear. The southerners were opposed to any royalty because they said jukeboxes were the 'cultural mainstay of the South.' I though that this was somewhat demeaning to their region, but nonetheless, that was their argument. And so the bill would disappear, and in the next Congress the whole process would have to be started all over again, and it went on year after year after year, and wasn't resolved until the 1976 general revision.

"To me this is an enormously important intellectual problem for the future," Mathias went on, with passion in his voice. "If we're going to deal more and more in a world of ideas—in a service economy instead of a productive economy, as the economists say—we have to find a way to protect them. We're working on a method of protecting computer software, which is not an easy thing from a technical point of view. Other nations have been able to do very well in—well, I won't denigrate their efforts by saying 'imitating our computers,' although in many cases that is what they have blatantly done—but in taking the computer concept and developing it on their own. Software really is the place where the United States has not been seriously challenged up to this time." It struck Mathias as ironic that a number of European countries, including West Germany, Hungary, and Austria, had adopted royalties or taxes on VCRs and/or videotapes, much as the MPAA had advocated (although the proceeds tended to go to public support for the arts rather than to individual copyright owners). By its failure to act, Mathias feared, the United States was putting the brakes on what might otherwise be a powerful international trend. "Unless we can protect intellectual property," he said, "then a lot of what we are going to be creating will be plagiarized or pirated or trespassed on, and that will mean an awful lot in plain dollars-and-cents terms to Americans in the future. I'm sure the video dealers who put up signs saying, 'Senator Mathias wants you to pay more for your videocassettes' don't think in those terms. Their concerns don't get into the highly philosophical."

One of his subcommittee staffers nodded and observed, "When people say, 'It's not the money, it's the principle of the thing,' you can be sure: It's the money."

Mathias took out a piece of paper, wrote something on it, and passed it across the table. He had drawn up a table:

Era of Monastic Manuscript	*Era of Gutenberg Press*	*Era of Promiscuous Publication*
Copyright unnecessary to authors or publishers	Copyright necessary to authors and publishers	Copyright enforcement doubtful

The waiter brought a trayful of mimosas—half orange juice and half champagne—and Mathias offered a toast.

"To copyright," he proclaimed.

Amid the tinkle of glass, half a dozen voices chimed in: "To copyright."

COMPLETELY SNOOKERED

AFTER THE BETAMAX decision the studio executives also met to talk about the future. Showing the same good judgment as the Argentinian generals who decided to put off the reconquest of the Falklands after their defeat by the British, Hollywood's generals agreed that 1984 was going to be a bad year for the pursuit of a home taping royalty. So they resolved to set that goal aside, temporarily. This was a difficult decision for Jack Valenti and Sidney Sheinberg; but in other corners of the moviemaking world there was a sense of relief that the slate had been cleared, and there was hope that the industry, freed of the burden of a notorious lawsuit, would fare better in Washington from there on out.

Several top studio executives and their lobbyists had long been persuaded that they represented a shrewder, more sophisticated, more forward-minded view of the world than their colleagues at Universal and the MPAA. "Universal is so incredibly hard-line," a lobbyist for one of the other studios commented, anonymously, after the decision. "Their reaction to this technology is: 'Let's squelch it!' And they have totally dominated the MPAA." Hollywood, he went on, should stop treating the VCR as an "unwanted intruder" and start thinking about how best to exploit it—and about what barriers, if any, stood in the way.

The home video business had entered a period of phenomenal growth, but there was still plenty of dissatisfaction with it in Hollywood. In 1983 the studios had realized about $400 million in revenues from home video, while the retailers were estimated to have grossed about $1.2 billion. In the theatrical movie business the studio people were used to coming away with about half the money paid in at the box office, and the one-third share they seemed to be getting in home video was disappointing. So was their continued inability to stimulate the so-called sell-through market. Paramount had tried, by pricing such movies as *Star Trek—the Wrath of Khan* and *An Officer and a Gentleman* at $39.95 retail (or about $30 wholesale). As a result of its new pricing policy, Paramount had sold many more cassettes than before; but its gains, in terms of dollars, were debatable, and there had been little increase in the number of cassettes the stores sold to the public. The main effect of Paramount's strategy, most people in the industry believed, was increased profit margins for the video dealers, who did the same thing with Paramount's titles they did with everyone else's: They rented them out.

The management of Twentieth Century-Fox continued to believe that the only solution was for Congress to change the law on videocassette rentals. The people at Disney were also high on the rental issue and low in their opinion of Universal's tough-guy attitude. Disney's executives believed that they were in an unusually good position to sell—as well as to rent—their products because children's programming had the attribute known in the home video world as repeatability. It followed that Disney had more to gain than most of the studios from a law that would bar retailers from renting out videocassettes without the copyright owner's approval. Disney and Fox had taken the lead the previous year in getting this idea split off into a bill of its own in both the House and the Senate. Now they intended to press for its passage.

Conversing among themselves, away from their mass meetings, some of the Copyrightists wondered if they wouldn't be better off with a more temperate spokesman than Jack Valenti. He had made some pithy contributions to the discourse, to be sure. A congressional hearing was hardly complete without a Valenti line like "If what you own cannot be protected, you own nothing," or without his description of the VCR as "an unleashed animal" or "a Pac-Man devouring everything in sight." But there were times when Valenti plunged into the sea of metaphor with a bravado that worried even his most ardent admirers. At one hearing he remarked that "the VCR is to the American film producer and the American public as the Boston Strangler is to the woman home alone." In an interview with *Variety* Valenti predicted that Americans would own fifty million VCRs by 1990, and "the minute they turn them on," he said, "there will be fifty million tapeworms ticking while the after-market withers and atrophies."

Bruce Lehman, Fox's lobbyist, knew that his former boss, Bob Kastenmeier, was tired of the movie industry's whining and did not respond well when, for example, Valenti portrayed copyright ownership as "just like your home or your car." For some months now Lehman had been trying to elevate the tone of the debate. At a House Hearing in October 1983 Lehman's present boss, Alan Hirschfield, the chairman of Fox, testified that the video rental bill, far from being a money grab by Hollywood, was a proposal that would facilitate "the shift from a smokestack economy to a prosperous technology-based society." Hirschfield couldn't match Valenti in the profile recognition department. His jaw was not so sharp; his hair, not so thick and silvery. But he had an air of celebrity as the protagonist of the best-selling book *Indecent Exposure,* a chronicle of his long and ultimately unsuccessful campaign, while president of Columbia Pictures, to fire the confessed check forger and embezzler David Begelman, and he came equipped with as many "new ideas" as Senator Gary Hart, who was running for president at the time. "I am not here to tell you that Hollywood is on the ropes," Hirschfield testified. "I am here to make the case that with the passage of this very simple amendment—which will take nothing out of the public or private domain—the Congress has an opportunity to stimulate the growth of a whole new industry in our country. . . ."

Sometimes, Hirschfield said, "our industry gets tarred with the brush of being 'those damned moguls with their Mercedes and their Cadillacs'—and there're too damn many of them around, I might add." But it was important to remember that the combined revenues of all the MPAA companies were less than five billion dollars—less than those of a medium-size corporation. Moviemaking was not a large industry ("excluding Dolly Parton," he jested); it was "a relatively small but unusually high-profile segment of the American economic community," which, with Congress's cooperation, had the potential to expand beyond entertainment into the booming field of information. It could become "a dynamic and vibrant video software business"—Hirschfield did not flinch from using a word that Valenti abhorred—"existing side by side with a dynamic and vibrant video hardware business.

"We want nothing at the expense of other segments of the home video community," Hirschfield insisted. "We only desire to stimulate the growth of the entire industry." Retailers and producers alike would profit, his argument went, if prerecorded cassettes could be offered to the public at dramatically reduced prices. If the rental bill passed, Fox promised to go on making all its properties available for rental and not to try to force rental prices up. The one thing it would do differently was to create a new, less expensive category of cassettes—"to be sold for private use only." Hoping to bridge the grievances that had divided the video dealers and the studios in the past, Hirschfield offered his "public pledge" that Fox would work to develop a marketing system that "harms neither the retailer nor the consumer."

He showed equal sensitivity to the concerns of Chairman Kastenmeier—or of any liberal legislator with reservations about the property right thrust of Valenti's rhetoric. Copyright, Hirschfield told the House, "was never intended as a band-aid to protect outmoded industries; nor was it conceived of as some kind of inherent right. It is a legislative tool, mandated by the Founding Fathers, to stimulate new creative enterprises—exactly the kind of enterprises our MPAA companies will develop if the law is updated with the modest changes proposed in H.R. 102." In conclusion, Hirschfield complimented Kastenmeier on his leadership in getting the 1976 Copyright Act passed, "and on a more personal level," he added, "I'd like to say that I don't think Wisconsin should be a twelve-and-a-half point underdog to the Gators this weekend."

The people at Fox and Disney hoped the rental issue would be less controversial than the royalty. The repeated rental of a videocassette, it seemed to them, was no different in concept from a public showing of a movie in a theater, and the law ought to treat the two activities the same way—as commercial uses of a copyrighted work, to be allowed or disallowed at the copyright owner's discretion. Politically the moment seemed right. Hollywood's friends in Congress were bubbling over with regret about the unfortunate resolution of the Betamax case and the unfortunate fact that it would not be possible, in an election year, to impose a tax—as it would inevitably be called—on home tapers. The rental bill was a relatively obscure and technical affair which hardly anybody but a copyright scholar could understand. It was a favor that Congress, cognizant of the support many of its members had received from the movie industry in the past, ought to be able to perform.

At an earlier period in the Betamax dispute Jack Valenti had described the Home Recorders' opposition to the rental bill as "a little Machiavellian trick that the EIA is trying to pull." Wayman and the EIA were using the video dealers as "a political bargaining chip for the copyright royalty issue," Valenti said. Sooner or later, he was sure, an emissary from the other side would offer to concede the rental issue if the movie industry would abandon its quest for a royalty, and when that happened, Valenti added, the bearer of the offer would "get promptly thrown out of my office."

These expectations were based on accurate, but somwhat outdated, intelligence. The Home Recorders had given some thought to the possibility of a trade-off, but sometimes a policy is adopted for one reason and continued for another. Six months before the Supreme Court's decision Dennis DeConcini, the Home Recorders' most persistent advocate in Congress, had given them an ultimatum. DeConcini and his legislative director, Romano Romani, felt that the senator had received pitifully little support or publicity for his efforts on behalf of the electronics industry. DeConcini had been forced to cancel a fund-raising party in Hollywood because the host, a lawyer with movie industry connections, had discovered high levels of Betamax-related hostility to the senator among potential invitees. (He had gone ahead with two other L.A. fund-raisers, however, "and Jack Valenti contributed to my campaign even in the heat of all this," DeConcini recalled, "so it looked to be far more minus than it turned out to be.") Movie people, clearly, know of DeConcini's opposition to the royalty; it seemed to him that the beneficiaries of his stand ought to be equally aware of it and ought to do more to show how much they cared about the issue. "A lot of them," Romani complained, "just want to roll over and play dead." Anticipating a more serious effort to pass the rental bill, Romani let the lobbyists know that they had better mount an aggressive grass-roots effort against the bill if they wanted DeConcini to spearhead the opposition this time as he had on the royalty. Otherwise, they could find another senator to speak for them.

DeConcini's ultimatum forced the Home Recorders to look hard at the rental question. A few of them had been drawn to it from the beginning. Carol Tucker Foreman, who had been retained by the coalition to line up the support of the consumer movement, saw the Hollywood bill as a case of attempted price-fixing. The legislation, christened the Fair Marketing Amendment, gave the studios the power to bar rentals if they chose, and Foreman reasoned, "They could say, 'From now on you may not rent this tape for less than five dollars a night.' That's retail price maintenance," she continued, "and there's nothing designed to get a consumer advocate's adrenaline flowing faster. It's the original bugaboo. It goes back forty years. That was *the* issue that kept the consumer movement going until the days of Ralph Nader and health and safety." To a consumer advocate, Foreman added, the rental question was "much more fun" than home taping. It resonated just as strongly with Nancy Buc, Matsushita's representative on the coalition's steering committee. Buc was a former antitrust lawyer, and she saw the rental question in antitrust terms. By selectively giving and withholding the power to rent, Buc argued, the studios could decide which retailers they wanted to do business with and which they didn't, and on what terms.

Foreman and Buc were the exceptions, however. By and large, the Home Recorders felt that the rental issue, as a software question, was not a proper concern for a coalition of hardware companies. Gary Shapiro, the general counsel of the EIA's consumer electronics group, had been allowed to engage in a modest amount of organizing work with the video dealers, but the coalition was not about to commit serious money to the kind of undertaking that DeConcini and Romani had called for. Sony, which had borne the brunt of the lobbying burden at the beginning, was cutting back on its financial participation because of declining profits, and the lobbyists for some of the other companies seemed to be suffering from serious battle fatigue.

But David Rubenstein and Ron Brown, the two lobbyists who had been retained by the Electronic Industry Association of Japan, saw things differently. VCR sales in the United States were increasing at a fast clip—800,000 units in 1980, 1.4 million in 1981, 2 million in 1982, and a projected 4 million in 1983—and buyers were starting to cite the ability to rent movies as one of the product's main attractions. At first the availability of the hardware had spurred the demand for software; now the process was working the other way, too. If the law gave the movie industry the power to control rentals, many video dealers feared that Hollywood would use the leverage to raise rental rates as part of a campaign to encourage consumers to buy. If that happened, fewer people would care to own VCRs in the first place. It followed that the manufacturers had a stake in the first sale doctrine—the provision of the law that allowed retailers to rent out their cassettes without the studios' permission.

Or so, at any rate, Rubenstein and Brown advised their Japanese clients, and their advice was persuasive. After making a formal presentation of their case in Tokyo that August, Rubenstein and Brown returned to Washington half a million dollars richer, and they began spending that money hiring organizers to fly across the country and alert video dealers to the perils of the rental legislation. The campaign began in August on a trial basis, with Tucson and Phoenix chosen as the first two stops, largely in order to "show DeConcini that we were behind him—that he wasn't alone in the wilderness," as Jeffrey Cunard, one of Sony's lawyers, recalled. As the guest speaker at these gatherings Cunard explained the legislation to an assemblage of local video dealers and recorded their response on what he called "a video petition," which was later submitted to DeConcini and other Arizona Legislators. "We thought it was a nice use of the technology," Cunard said, "and one could well ask, 'Should that tape be taxed?' "

Anger is always good for a headline. The dealers who came to these meetings were angry. "If Hollywood had its way, we would be at their mercy, and they don't take prisoners," one dealer would say. Another would tell about having mortgaged his home in order to launch his business, and ask, "What protection do we have if they say, 'Take a long walk off a short pier'?"

From where the video dealers sat, it looked as if Hollywood was doing just fine with the law the way it was. "If the chairman's not showing losses to the stockholders, I don't have much sympathy for the complaint that he needs more," Frank Barnako, the incumbent president of the Video Software Dealers Association, observed. "It's the movie industry guys at the hearings who are wearing the green-and-red-striped moccasins. My guys are wearing double-knits. I don't know anybody in video who drives a Mercedes. I can't feel sorry for Charlton Heston and Clint Eastwood."

Most of the video dealers thought that Hollywood ought to be thanking them—and the electronics industry—for a new source of revenue, not only on current movies but on "a lot of movies that were lying around in a vault somewhere," as Ira Gomberg observed.

"This is all gravy money" was how George Atkinson, the president of the Video Station in Los Angeles, put it. "They're saying, 'I'm not happy with the gravy. I want ice cream on my gravy.' "

Accurate or not, this was a view of the issue which local newspapers and TV stations found worthy of coverage. "We'd talk to people at the meeting, and then I'd go back to my hotel room and watch myself on the evening news," Cunard recalled. "Often times it would be a pretty major story in places like Greenville-Spartanberg, South Carolina. And there isn't that much to cover in Mobile."

The meetings had educational value for the lobbyists as well as the dealers. They learned, for example, that they got better coverage if a dealer alerted the press than if a hired PR person did it. And it made a better impression to have the main speech delivered by a dealer than by a Washington lawyer like Cunard who could be depicted as an outside agitator.

Encouraged by the response, Rubenstein hired a veteran of the video software business, Risa Solomon, to serve as the full-time organizer of the grass-roots campaign. Solomon was a red-headed Texan who had worked for a Dallas video store and had served as the first executive director of the Video Software Dealers Association. Her commitment to home video was personal as well as professional. On Saturday mornings in Dallas, where she lived, the TV schedule was dominated by kiddie cartoons, and on Sunday mornings by fundamentalist religion. Home video had given Solomon a wider choice over what her children watched. "If they have seen *Mary Poppins* and *Sound of Music* less than fifty times each," she said, "I'll eat my cassettes."

In one month—November 1983—she hit Las Vegas; Beaumont, Texas; New Haven; Salt Lake City; Cincinnati; Chicago; and Grand Rapids. She suspended her trips for the month of December because "dealers won't leave their stores during the Christmas season," she explained. But she was back at it in January 1984, touching down in Wichita, Kansas; Tulsa, Oklahoma; Des Moines, Iowa; Raleigh, North Carolina; Louisville, Kentucky; and Fort Lauderdale, Florida.

The typical meeting was held at a convenient hotel in the early evening of a Tuesday or Wednesday—the slow days of the week in the video business—and the bar was open. Without fail, the dealers were urged to seek an appointment with their senators or representatives. As David Rubenstein explained, "Our view was that a face-to-face conversation between a member of Congress and twenty or thirty dealers, telling them why their business was going down the tubes if this Hollywood-inspired legislation were to pass, and why nobody from their state was going to benefit from it, would be very useful."

Not all members of Congress were eager to grant these audiences, but few could remain unavailable for long in the face of a sustained barrage of mailgrams signed by hundreds of their constituents. And the Home Recorders made it extremely easy for a dealer to get his name on a Mailgram; all he had to do was call a toll-free twenty-four-hour-a-day, seven-day-a-week phone number, and his signature would be added to the rest. In response to one such communication, Senator Joseph Biden of Delaware agreed to talk to a group of dealers over a telephone conference line. He spent an hour and fifteen minutes on the phone with them, and an eavesdropping lobbyist heard him say, "Let me tell you guys up front—you've got all the political arguments. I know if I vote against you, there'll be signs in the stores saying, 'Joe Biden wants to make you pay more for your videocassettes.' " At the end of the call Biden asked one of the Delaware dealers to hold a copy of *Star Wars* for him to rent when he got home for the weekend.

The modern lobbyist will acknowledge that his profession, high up the evolutionary ladder though it may have climbed, still harbors a few prehistoric relics known as door openers, who eke out a living selling access to key government officials with whom they happen to enjoy a close relationship. Usually the door opener will be described in the abstract, and he may be understood as one who, like the fly-by-night used-car dealer, provides a contrast for the more respectable members of his calling to illuminate their high standards against. But there are times when a living, breathing, identifiable door opener crawls out from under his rock and into plain view. Such a time came in the summer of 1983, when the movie studios set out to win the Reagan administration's backing for their rental bill. Unsatisfied with the MPAA's White House connections, Twentieth Century-Fox decided to retain a lobbyist who could command the attention of the President and his closest advisers. Fox chose Lyn Nofziger, a longtime member of the Reagan inner circle who had left the White House staff in order to advise the president at the higher rates of pay available on the outside.

When the Home Recorders realized what the opposition was up to, they looked around for a door opener of their own. First they went after John Laxalt, the brother of the Nevada senator, Paul, who was one of the president's closest political allies. But Laxalt's terms were stiff—"something like a guaranteed contract for two years at twenty-five thousand dollars a month," according to one of the lobbyists who dealt with him—and he could not be bargained down to a figure the Home Recorders were prepared to pay. Eventually they settled for Peter Hanaford, a public relations man and former partner of Michael Deaver, the president's deputy chief of staff. Hanaford dutifully arranged a meeting between Hall Northcott, a lobbyist for 3M and a charter member of the Home Recorders, and Edwin Meese, with whom the coalition had unsuccessfully sought an appointment for months. "It was a cordial meeting," Northcott recalled. "He had not spent a lot of time studying the issue; but he clearly picked it up very quickly, and he understood our perspective. He gave no commitments. At the same time he promised that if he heard of anything happening in this issue area, he would let us know."

In October the White House let the Home Recorders know that it would be siding with the movie industry. A series of administration witnesses were dispatched to testify in favor of the rental bill before the House and Senate. The studios, according to *TV Digest,* were showing the "kind of lobbying power for which they're famous."

With the Betamax case out of the way, the Hollywood forces moved quickly to get some action on the rental bill. Two weeks after the Supreme Court had ruled, Senator Mathias's office quietly scheduled a markup session for February 22, and the Kastenmeier subcommittee announced that it would hold a hearing on the rental issue the following day. The Home Recorders assumed that Mathias would not be taking this step unless he had rounded up the necessary four votes on his subcommittee. Three members (Mathias, Dole, and Orrin Hatch) had already declared themselves in favor, and only one (DeConcini) was known to be against. It was hard to imagine that the three remaining members (Leahy, Laxalt, and Howard Metzenbaum) would not provide at least one more vote. Laxalt took the same side as the administration on most issues, and he was also close to Hatch, a fellow conservative from Nevada's neighbor to the east, Utah. And while the Home Recorders had tried and failed to hire his brother as a lobbyist, the MPAA had succeeded in retaining Laxalt's daughter, Michelle, to lobby for the rental bill. It was hard to imagine a door opener with better access than that, and the bill's opponents were not much reassured by Michelle Laxalt's statement that as a matter of principle, she would not lobby her father. "I don't touch Laxalt," she explained. "That's been my modus operandi as long as I've been in this town."

The Home Recorders' confidence was further shaken by the news that Romano Romani, who tended to know which way the wind was blowing, had resigned from DeConcini's staff in order to form a legislative consulting firm ("It's our euphemism for lobbying," he explained) with the MPAA as a charter client. Romani hadn't made his move without warning. "He called us in December and said he was going to form a firm with Tom Parry from Hatch's staff," one of the Home Recorders said later. "He wanted to know if we had any ideas for potential clients. Obviously he wanted to be hired by our coalition, and some of us recommended that we hire him because we figured he had good insights into the Senate Judiciary Committee and he knew the issue and, I mean, how much more is it going to cost?" While the Home Recorders were dickering about the value of Romani's services, however, he called back to say that the MPAA had made him an offer he couldn't refuse—unless, perhaps, he heard a better one. Offended by the idea of bidding for the services of someone who was threatening to go over to the enemy, the coalition decided to let Romani defect. Nevertheless, he was able to strike a deal with Sony to represent it on tax and trade matters even as he represented the MPAA on the rental issue. From his work on the Betamax issue as a Senate staffer, Romani had snagged his first two clients as a lobbyist.

On February 4, a Saturday, the Home Recorders called an emergency meeting to assess the situation, and they resolved to do what they could in the week and a half remaining before the markup session. By now they had come to value the rental issue for its own sake. Their alliance with the video dealers was a blind date which had panned out. As Hall Northcott of 3M said, "We wouldn't trade that issue for love or money."

Mailgrams and phone calls went out to more than fifteen hundred video dealers in the seven states represented on the subcommittee, informing them of the need to telephone their senators and urging them to pass the word to their customers. A supply of flyers was shipped out, and the coalition bought thirty thousand dollars' worth of newspaper ads in the seven states. A half-page ad in the *Baltimore Sun* asked readers, "DO YOU WANT TO PAY MORE TO RENT MOVIES? IF NOT, CALL SENATOR MATHIAS TODAY." The public was warned that "if Hollywood has its way," customers would pay higher rental prices and have fewer titles to choose from, while "thousands of small neighborhood stores" would be driven out of business. "Why is Hollywood pushing S. 33?" asked the ad. "The studios want Congress to give them a 'consolation prize' for having failed in the Supreme Court to gain control over the use of your home video recorder." Follow-up ads ("Senate votes tomorrow! Call today!") appeared as the markup drew nearer, and some dealers put telephone auto-dialing machines on their counters so that customers could contact their senators right then and there, without taxing their finger muscles.

The result was a ferocious number of phone calls. At one point

Senator Dole's office in Kansas City was said to be getting a thousand calls an hour. Some of the staff members who fielded these calls had to apologize for the fact that they had never heard of the bill in question. "It was really way down on their list of priorities," Frank Barnako said. But the Home Recorders were not going to rely exclusively on long-distance artillery. Having caught the senators' attention, they recruited delegations of video dealers to fly to Washington for some last-minute lobbying, armed with petitions with more than a hundred thousand signatures.

On Tuesday, February 21, the day before the markup, some twenty video dealers—several from every state with a senator on the subcommittee—flew into Washington and checked into the Ramada Renaissance Hotel, their transportation and rooms paid for by the EIAJ. At a luncheon briefing session David Rubenstein welcomed them and said a few words about how they might make themselves most effective. Risa Solomon cautioned them not to "act like idiots and warmakers. You're professionals," she said. "You're explaining issues and what's at stake." Jeffrey Cunard explained the bill and gave a summary of the best arguments to be made against it—among them, that it would encourage price-fixing since the studio might say, "From now on you may not rent this tape for less than five dollars a night," and that it would enable the studios to determine who would or wouldn't be allowed into the home video business. Gary Slayman, one of Matsushita's lawyers, outlined some reassuring amendments that the Hollywood forces were expected to introduce and warned the dealers not to be drawn into discussions of these amendments lest they send a "false signal" to the senators that the bill might be acceptable to them with certain changes. Finally, Ron Brown gave a short speech that was a "combination of 'How to Talk to Your Senator' and a peptalk," according to one lobbyist, who added, "It was very effective. He got me psyched up."

When the dealers arrived on the Hill that afternoon, a group from Kansas met with Doug Comer, an aide to Senator Dole. Alan Anton, the co-owner with his father and brother of House of Anton Video, a chain with stores in Oberlin Park, Kansas, and Independence and Gladstone, Missouri, recalled that Comer made "a very noble-type statement" about Dole's position, "with no details whatsoever." The visitors asked if they might see the senator in person, and "we were told that he was out of town and due back in, but nobody knew just when. We found it a little dismaying that nobody knew what Senator Dole's schedule was," Anton added. "We thought that maybe they were trying to sidestep us."

Even Dole's personal secretary professed ignorance, but she told the visitors to call back later, just in case. Instead, they dropped in without warning at five o'clock, and no sooner had they appeared than "we caught a glimpse of him walking in the back part of the office," Anton said, "and eye contact was made such that he was aware that we were aware that he was there." When the dealers reinstated their request for an appointment, it was promptly granted. Dole "gave us several minutes of his time," Anton said, and after hearing them out, he expressed his intention to vote in favor of the bill at the markup. He hedged some, however, by saying that "his feeling was that the full committee should look at the issue in fairness to all concerned." Thus he left open the possibility of voting differently later on. He added that he would prefer for the parties to resolve their dispute without Congress's help.

A group of Utah dealers staged a sit-in in Orrin Hatch's office and vowed to remain until the senator agreed to see them—as eventually he did. When Dale Snape learned about the sit-in, he felt a perverse admiration for the video dealers' spunk. "If you're a one-issue group and this is the only thing you need to see your senator about," he explained, "you can afford to do things that a lobbying organization like us wouldn't be able to do."

After a rather stately and inconclusive meeting with another senator a dealer deliberately forgot her pocketbook in order to have an excuse to return and find out what he was really thinking. "That's not something you teach someone," Joseph Cohen, the executive director of the VSDA, observed.

The night before the markup a group of lobbyists and dealers compared notes over dinner at the Sichuan Pavilion, a fashionable Chinese restaurant on K Street. There was good news from the Nevada dealers, who thought they had made a favorable impression on Laxalt. But the

Home Recorders still had only one commitment—DeConcini's. After dinner David Rubenstein stopped by the *Washington Post* building on Fifteenth Street to pick up a copy of the next morning's newspaper, which he eagerly opened to the editorial page. Rubenstein had a number of friends in the press—people who relied on him for tips and unattributed quotes. In the last few days he had spent quite a bit of time on the phone with a *Post* editorial writer, trying to explain the background of the rental issue. Naturally he was curious to see the result, and it pleased him. Under the headline "Hollywood and the Free Market," the *Post* declared that the rental bill would allow the studios to "squeeze small dealers out of business," although it was they "who had the foresight to recognize, as Hollywood did not, that rentals, not sales, were the main market, and who ran considerable risk in stocking their extensive tape inventories." The *Post* concluded that "the studios need to build a better case that they can't prosper in the current market before Congress decides that Hollywood needs a helping hand."

Rubenstein was equally interested in another item in the *Post*—a story in the financial section headlined "Hollywood Lobby Blitzes Hill." He had played a part here, too. "In a high-budget sequel to last year's battle of the Betamax, the Hollywood film industry has launched an intense lobbying blitz for legislation that would gurantee them a share of the burgeoning market in rented home video movies," the story began. "Led by Jack Valenti, president of the Motion Picture Association of America, the Hollywood forces have beefed up their already star-studded cast of lobbyists by hiring the daughter of Sen. Paul Laxalt (R-Ne.), and a former top aide to Sen Dennis DeConcini (D-Ariz.) who acknowledges that he has spent at least some of his time making calls from DeConcini's office." The reporter, Michael Isikoff, went on to reveal that Romano Romani had been "reached twice by phone yesterday at DeConcini's Senate office."

The article could hardly have come out any better if Rubenstein had written it himself—except for one minor point. Normally he was careful to cover his tracks by asking reporters to describe him as "a Washington lobbyist." This time he had foolishly acceded to the reporter's request to have one quote on the record. Responding to the movie industry's claim that it wanted the rental bill in order to lower sales prices, not raise them, Rubenstein had said, "Nobody in this city goes up to Congress to lobby for a bill that's going to lower prices. It would be the first time in the history of the republic." Since he was the only anti-Hollywood lobbyist named in the article, the educated reader would have no trouble fingering him as the principal source of the damaging information. Rubenstein was embarrassed, but not as embarrassed as Romani and the MPAA would be.

The story had not been planted out of pique, however. Rubenstein's purpose was to send a message to any wavering senators who might have notions of voting Hollywood's way. In particular, he was after some insurance against a last-minute switch by Senator DeConcini, who had supported the movie industry in the past and, it was feared, might wish to return to its good graces.

By nine-thirty the next morning dozens of lobbyists and video dealers had collected in a corridor outside the hearing room where the markup was to be held. Risa Solomon, who had done much of the phoning and stage managing of the dealers, was excited. "I couldn't sleep all night," she said. "I can't believe how tenacious this group is."

"Dole said that his people have never seen as much mail on a consumer issue as this one," Joe Cohen of the VSDA reported.

"A week ago there could have been almost a unanimous vote against us," said Jack Messer, a dealer who had flown in from Cincinnati. "Now it looks like it's going to be close."

Just before ten the word began to circulate: The other side had come up a vote short, and Laxalt was the holdout. Hatch, the story was, had pleaded with Laxalt to support the bill, but in vain. Events inside the hearing room were in keeping with the rumor. Only Mathias and Laxalt showed up, and Mathias briskly announced to the reporters, video dealers, and lobbyists in attendance: "I have asked the staff to canvass the absent members. It would appear from the reports that we are getting that we will not attain a quorum this morning." At this the throng of dealers applauded, rather rowdily. "So we will reschedule this hearing for the earliest convenient date," Mathias continued in a soft voice. "Good morning." And the markup session disbanded, with no legislation marked up.

As the crowd filtered out of the hearing room, several of the Copyrightists could be heard trying to persuade reporters that this was a temporary setback. "Don't let appearances deceive you," Michael Berman, Columbia Pictures' man in Washington, was saying. But the press, by and large, failed to heed his warning. The *Washington Post* reported the next day that the subcommittee, "bombarded with thousands of phone calls and letters from video cassette dealers and their customers, yesterday temporarily dropped its efforts to give Hollywood film producers a share of the proceeds in rented video movies." The *Post* quoted one of the victorious lobbyists as saying, "We had dealers coming out of the walls," and vowing, "We're going to keep bringing the dealers back." A few days later Senator Dole dropped his name off the rental bill as a cosponsor. "Big Mo appears to have changed direction" was the way *Video Week* summed up the developments.

A few hours after the aborted markup, Fox's lobbyist, Bruce Lehman, walked into a French restaurant near his office in Georgetown, sat down, and ordered a drink. He was distressed, he said, about the failure to protect America's intellectual-property interests. "I think one of our greatest national assets is just being frittered away," he explained. "I suppose I shouldn't care about this as strongly as I do. I should just laugh all the way to the bank." The Home Recorders had surprised even themselves, in Lehman's opinion, with the strength of their grass-roots campaign, and success had made them unyielding. "Their opposition has taken on a life of its own," he said. His side, he added, had made serious tactical errors—above all, the initial one of putting the royalty and the rental proposals into the same bill. "What that did," Lehman explained, "was to make the people in Tokyo and their hired guns say, 'Here's another issue we can exploit. If we fire up the video software dealers, we'll have a grass-roots base.' We walked right into that trap with a decision that was made at a very low level, without much thought."

Dale Snape was also in a downbeat mood when he sat down for an interview in his office later that afternoon. He wondered if it had been "a personal failing on my part that I cannot seem to bring people to the conclusion that the consumer's stake comes before the time when he is paying for something—that it comes when choices are being made about what he can choose among to buy. It is personally very sad for me," he said, "that groups like the Consumer Federation of America and Consumers Union look at these proposals in a hostile way." He was in awe of the electronics industry's ability to persuade people that "because they sell to consumers, they ipso facto represent consumers. The leap of faith you have to make," he said, "is just staggering! The Consumer Product Safety Commission *exists* because of concern about manufacturers. I cannot tell you the frustration that that causes me, personally and intellectually."

Snape had had the unhappy assignment of marshaling consumer support for the MPAA—a task that had matched him up against Carol Tucker Foreman, working the same beat for the Home Recorders. The high point of Snape's campaign had been the time he persuaded the Consumer Federation of America *not* to take a position on video rentals at its 1982 convention. A year later, when the association's executive committee formally approved a resolution opposing the MPAA's bill, Foreman, a former president of the CFA, was in the room while Snape was asked to leave. "If you're going to organize consumers, hire Carol Tucker Foreman," said Snape by way of summing up that phase of the battle.

He was trying not to dwell on the past, however. "We'll be looking in the future to a broader coalition that won't be founded so much on motion pictures," Snape said, "because certain copyright holders are more attractive than others, and I would say as a class the motion picture studios are the least attractive in the public mind. I will be recommending a non-Betamax-type press and public relations campaign with its focus on educating people on the value of copyright and the reprographic revolution. Not the kind of thing you can whip together and expect to get passed in this Congress. A public awareness type of thing." He was in the process of preparing a written proposal for submission to the MPAA.

In the course of the following year the Ninety-eighth Congress passed several intellectual-property measures, including an audio rental bill supported by the recording industry and a bill that gave quasi copyright protection to computer chips. No further action was taken, however, on either of the MPAA's proposals, and no one even bothered to introduce them when the Ninety-ninth Congress convened in January 1985. The recording industry expressed renewed determination to pursue an audiotaping royalty that year, but this time the audio people would go it alone, unencumbered by the political burden of an alliance with the movie industry.

Jack Valenti professed to be hopeful that Congress would, in time, come around. At the moment he had other worries. MPAA surveys indicated that a growing number of people were habitually copying rented movies. This was an offense that, for the moment, required the use of two VCRs, but a Japanese company, Sharp, had just made the job easier by creating a double-slotted VCR which served as a videocassette-copying machine. Valenti put little faith in Sharp's assurances that this "two-headed dragon" or "amoeba run amok," as he called it, was not intended for export to the United States. "I don't believe that for one moment," Valenti said. "If they think there's a market for it, they'll do it. But that's just the Conestoga wagon of the jet plane in this whole business." Vidicraft, a small company based in Portland, Oregon, had come out with a product called the Commercial Cutter, which automatically removed commercials in the act of videotaping (by detecting the breaks between segments of a TV broadcast, timing each segment, and retroactively erasing those that lasted thirty seconds, sixty seconds, or any of the other standard commercial lengths). Valenti had predicted such a device—and the Home Recorders had scoffed. The next thing coming down the pike, he feared, was a product that would allow people to copy a videocassette in a fraction of its playing time. "The Japanese are working on it," he said. "I'm told that they have methods now of copying a two-hour movie in ten minutes. They're going to avalanche real time."

Looking back on the collapse of the movie industry's lobbying efforts, Valenti agreed with Dale Snape's assessment. "You've got to be proconsumer, and we were never able to show that we were proconsumer," he said. He singled out the press as another big headache. The *New York Times,* the *Wall Street Journal,* and the *Washington Post* had come out against the rental bill, as they had against the royalty. "We got completely snookered—just absolutely destroyed—in the press!" Valenti exclaimed. "Suppose I took the *New York Times* and put it on television—on some public access channel—and there was a machine that would allow you to copy it while you sleep and you'd have your own copy in the morning, without the ads. Would the *New York Times* find that to the public's benefit? You bet your ass they wouldn't!" He lowered his voice to a meditative timbre and went on: "But I suppose we lost this battle when those cartoons appeared about video police coming into your homes. The cartoonists killed us. We became objects of ridicule. Molière once wrote that most men don't object to being called wicked, but all men mightily object to being made ridiculous, and the Molière line is quite relevant today."

PIONEERING TAX

APART FROM A bare majority of the Supreme Court, hardly anybody of consequence had a good word to say for Sony in 1983 or 1984. The Betamax and the Beta format continued to lose ground in the United States, Europe, and Japan, in a trend that seemed stunningly insensitive to all of Sony's efforts to arrest it. The rise of the prerecorded cassette had made the shift from Beta to VHS a self-sustaining phenomenon: Because more people owned VHS machines than Beta, the video stores stocked more VHS cassettes than Beta; because VHS cassettes were more plentiful, new customers bought VHS machines. As one of the account executives working on the Betamax advertising campaign commented, it was a "killer problem."

By 1983 Sony's share of the world VCR market was down to twelve percent, and the company had been forced to cut prices sharply to avoid losing even more business. The result was an alarming decline in profits: from $307 million in 1981 to $186 million in 1982, to $127 million in 1983, a year in which Sony froze managers' wages and suspended bonuses for executives. And the company had an image problem that transcended its finances.

"Not long ago, the Sony Corporation seemed to be the ultimate consumer electronics company and the darling of Wall Street," a *New*

York Times article observed in August 1983. ". . . But a bad economy, tough competition on its home turf and a misreading of the video cassette market has changed all that . . . . Despite Sony's heritage, some analysts maintain that its present difficulties have done permanent damage to its future competitiveness, especially in the United States." Those same "some analysts," according to the *Times,* had begun "to wonder whether Sony is still a growth company," and one flatly (although anonymously) declared, "It's over for Sony. The company's best days are behind it."

Forbes magazine weighed in with an assessment of Sony's problems in October 1983. "The bold spirit of innovation that made Sony successful is no longer enough," *Forbes*'s correspondent, Michael Cieply, wrote. "A great company is in deep crisis." In times past, Cieply noted, Sony had been able to count on a two- to four-year lead in technology, but with its latest innovations—such as the Walkman personal stereo—the window of exclusivity had narrowed to between four and eight months. "Sony is producing premium products for a market where it can no longer charge a premium price," Cieply wrote. "With most of its innovations, it simply doesn't have time. . . . Sony pledges to be first with the best . . . but being first no longer carries much of a premium." Too often, according to Cieply and other critics, Sony's competitors were coming up with ways to make a Sony product more attractive to consumers, either by adding a few features—the bells-and-whistles approach—or by cutting corners to produce a cheaper facsimile. Sony, according to Ben Landis of the securities firm of Birr Wilson, had a tendency to "strive for more excellence than the market really demands." A phrase was coined to describe the position in which Sony kept finding itself: It was paying a "pioneering tax."

In the 1950s Sony had managed to maintain a monopoly over the Japanese tape recorder business by vigorously asserting its patent rights. By the eighties consumer electronics had become such a fiercely competitive business and the products themselves had become so fantastically complicated that no company could produce state-of-the-art equipment by relying on its own R&D. The sharing and trading of patents had become essential and routine, and when a truly new product emerged, its discoverers knew that no matter how secure their legal position, if they tried to construct a patent barrier against would-be imitators, the people they gave the cold shoulder to today could be the people they needed a favor from tomorrow.

Traumatized by the bad press, Sony struck back in January 1984 with a bold series of advertisements in Japanese newspapers. The ads pointed out that the Beta format still enjoyed a thirty percent share of the world market (it held a predominant position in South Africa, South America, and most of the Orient except for Japan); that the absolute number of Betamaxes sold was still increasing from year to year, despite the declining market share; and that various new features were making the Betamax "even more exciting." Unfortunately for Sony, these facts emerged in the course of an ad campaign that commenced with the question, "WILL THE BETAMAX DISAPPEAR?" illustrated with a cartoon depicting a pile of discarded Betamaxes in a garbage collector's cart. The idea was to grab the reader's attention and then set him straight. But as the journalist Masaaki Satoh commented, "It appeared to many people that Sony was making the case against the Betamax more strongly than the case for it. The ads ended on an upbeat note, but in the process of getting there they did a lot of damage, crystallizing doubts in the minds of consumers. In the history of Japanese advertising there is probably no other case in which the aim and the result have been so disastrously far apart."

Hot on the heels of the "WILL THE BETAMAX DISAPPEAR?" ads came Sony's annual shareholders' meeting. In Japan these affairs tend to be more spontaneous than in the United States, and sometimes they are enlivened by the presence of professional hecklers, known as *sokaiya.* Because of a change in the law barring companies from buying off the *sokaiya* with free meals and other favors, a customary management tactic, there were a number of exceedingly long and loud shareholders' meetings in 1983 and 1984, and none longer or louder than the one held by Sony on January 30, 1984.

"You're in the red, so don't act so big!" one jeering stockholder shouted.

"We're not in the red," Sony's president, Norio Ohga, replied. "Profits are down, but we will pay the same dividend as last year."

Another shareholder accused Sony of hiding its problems. "In the annual statement nothing is written about the decline in profits," he complained. "Only good things are discussed. And you make it out to be caused by the world recession and trade frictions—but other companies in the industry must have faced the same conditions. Didn't you rely excessively on your faith in the Sony brand? Didn't many of your management strategies fail?"

"I recognize our failures," Ohga said. "Because of them, we have maintained the dividend and given up the directors' bonuses."

"You think that's sufficient? Hah!" another stockholder retorted.

The criticism touched on every imaginable subject—even the unusually erect posture of President Ohga, a trait left over from his early training as an opera singer. The central theme, however, was the supposed imperial rule and high-flying style of Akio Morita. One angry stockholder suggested that the problems of the last few years could be traced to a "one-man show" style of management.

"That is absolutely not the case," Ohga said.

"That's what the outside directors are saying," the stockholder retorted.

"That is not the case," Ohga repeated. "I myself am inside the company, so I know the situation."

"Don't cover up for him!" someone shouted.

"I understand—I will watch out for that," said Ohga.

The meeting lasted thirteen hours in all—a new Japanese record—and as Kazuto Mimura noted in the book *Sony's Counterattack*, "the world took this as a symbol of hard times at Sony."

That spring Zenith became the last American OEM company to shift its allegiance from Beta to VHS, while three Beta manufacturers—Toshiba, Sanyo, and NEC—decided to hedge their bets by making VHS as well as Beta machines. Sony, as a matter of pride, had to defend its invention. The other companies, as a matter of common sense, had to go with the flow.

With the Beta family breaking up and the format headed toward a marginal existence, Sony plotted a new strategy. It would seek to establish Beta as the "high end" or semiprofessional VCR, while doing what it could to move the general consumer market from half-inch tape to eight millimeter, a format in which the cassette would be about the size of an audiocassette and the VCR itself could be built into the same unit as a camera to create a "camcorder" weighing only a few pounds. In January 1985 Sony announced two products designed to put the new strategy into effect. One was a "SuperBeta" VCR with a "significantly sharper, more detailed picture than ever before." The other was a "Video 8" camcorder, a one-piece camera/recorder unit which could be hooked up to a TV for playback. As Sony stated at every opportunity, it was "based on the common format" and thus "compatible with recorders of any manufacturer producing in this format."

Five companies—Sony, Matsushita, JVC, Hitachi, and Phillips—had participated in the initial talks leading to the eight-millimeter standard, which eventually was endorsed by 127 companies in all. Apart from Sony, though, none of the major Japanese manufacturers seemed to be in any hurry to market an eight-millimeter machine. Kodak, looking for a way to break into video, had made an OEM deal with Matsushita Electric for an eight-millimeter ensemble. But Matsushita had shown no interest in selling the same product under its own name, while a Hitachi executive had questioned the wisdom of "spotlighting a product before its time" and vowed that *his* company, for one, would not be switching formats "at an explosive growth-rate position in the industry's life cycle"—a time when "management should take responsible and prudent action not to interfere with the high sales rate currently being enjoyed."

With the move into eight millimeter, Sony returned to its customary position in the consumer electronics world—the vanguard. The stage seemed to be set for another technological leap forward, with Sony exploiting the appeal of home movies to carve out a place for eight millimeter, than gradually persuading the software companies to make prerecorded eight-millimeter cassettes, and finally building up consumer demand to a level at which even its most reluctant competitors would be forced to plunge in after. Then, it stood to reason, VHS would be the endangered format, and compatibility would become Sony's ally once again.

But this was a game at which two could play. Sony had based its new strategy on the assumption that home movies were the coming thing in video and that half-inch cassettes—either Beta *or* VHS—were simply too big for a home movie system. In May 1985 Sony came out with the Handycam, a three-pound camcorder which was far smaller than any Beta or VHS camcorder; in fact, it was about the same size as a VHS *cassette.* The Handycam got rave notices from the video columnists, and the move to eight millimeter seemed to be well under way. A year later, though, JVC came out with its Mini Video Movie, a camcorder scarcely larger than Sony's. The JVC system used VHS tape—but only an hour's worth—in a compact cassette which could be played back, with an adapter, through any VHS machine. JVC had a tremendous stake in the survival of VHS since it was not only a VHS manufacturer but the developer of the system and, accordingly, received royalties on VHS machines made by others. When the existing owners of VHS machines decided to buy a home movie setup, JVC was betting they would choose a more or less compatible product over an incompatible one.

In April 1986 six major Japanese electronics companies—Matsushita, Hitachi, Minolta, Mitsubishi, Sharp, and Toshiba—announced plans to sell camcorders using JVC's compact VHS cassette. They also announced their lack of plans to adopt eight millimeter in the foreseeable future. As *Television Digest* observed, "Picture has changed completely from a month ago, when virtually every VCR manufacturer was talking about adding 8mm camcorder. Difference has been extreme persuasion applied to VHS group by VHS inventor JVC. Result now almost looks like rerun of Beta-VHS battles, with Matsushita throwing its significant weight to JVC against Sony."

The "extreme persuasion" had been applied at a summit meeting of five leading VHS manufacturers, a group that (as the Japanese magazine *Nikkei Business* revealed) had been holding secret talks on a regular basis for nearly ten years. On March 7, they met at a hotel in Hamamatsu, a city midway between Tokyo and Osaka. Shizuo Takano, the senior managing director of JVC and a man sometimes called the father of VHS, greeted his guests by saying, "Get out of this room, any of you who want to start an eight-millimeter video business." No one left. Takano went on to say that the VHS format was under attack and needed defending. "VHS, however, retains a very strong advantage of compatibility among one hundred million units used in the world," he said. "JVC has developed a VHS compatible camcorder which can compete with eight-millimeter video in compactness and light weight. We provide this camcorder technology to you, our old comrades. Why don't you use our technology as a base and add your originality to see if you can expand your business?"

When word of the summit meeting got out, Sony officials charged that JVC had whipped its allies into line by threatening to withhold VHS patent licenses from any company that decided to sell eight-millimeter machines. Although the account in *Nikkei Business* failed to confirm Sony's charge, the video wars had clearly reached new heights of cunning and intensity. It was by no means clear, however, that JVC's strategy would prevail. At the slow, one-hour speed the picture quality of the VHS-C system was no match for eight millimeter at its normal two-hour speed, and at least two of JVC's allies seemed to be hedging their bets. Within a few weeks of the VHS summit meeting an impressive eight-millimeter camcorder—bearing the Olympus brand but made by Matsushita—was introduced at the summer 1986 Consumer Electronics Show. Hitachi, meanwhile, was reported to be making eight-millimeter machines for Minolta.

As for Sony, after watching one offspring get mauled in the marketplace, it seemed to be taking better care of its latest. At the end of 1986 the company launched an ad campaign which sought to acquaint American Christmas shoppers with the two types of camcorders. In a typical TV spot, a lawyer was summing up for the jury. "So the issue is clear," he declared. "Sony Video 8, the system of the future, or VHS-C, the compromise of the past." Not only did eight millimeter offer "a measurably brighter picture" and "far superior sound" but "on Sony videocassettes your memories are safe, while"—he pointed to a VHS-C cassette which had come hopelessly unraveled, and concluded in an ominous voice—"on VHS-C they could be fleeting."

Whatever Sony's eventual place in the camcorder market, the company seemed likely to endure. For all its problems it had somehow never reached the state of ruin widely forecast a few years earlier. Indeed, after the bad years of 1982 and 1983 its profits—and its reputation—had rebounded. Sony was being praised for "rationalizing" its purchasing and investment decisions; for offering a wider choice of features on certain products, such as radio cassette players; for building an assortment of models around a common chassis to hold down production costs; and for beginning to supply other manufacturers with such components as semiconductors and computer disc drives as part of a broad effort to reduce its dependence on the consumer market. And President Ohga, according to some observers, had taken to slouching his shoulders.

But a large share of Sony's revived prosperity could be attributed to the success of a new product, the mini compact disc player. An outgrowth of the same laser recording technology that had attracted MCA in the 1970s, the CD had begun as a Phillips invention and evolved into a Phillips/Sony collaboration. At eight hundred dollars and up, the original CD players had been slow sellers, and a number of companies, Matsushita among them, had lost interest in the technology. They got their interest back in 1984, when Sony came out with a product a twentieth as big and a third as expensive as the earlier machines. The new size had been determined by Kozo Ohsone, a Sony executive who had carved a block of wood into a piece about an inch and a half thick and five inches square—just larger than the disc itself. "To persuade the engineers," he said later, ". . . I told them we would not accept the question 'Why this size?' That was our size, and that was it." A *Wall Street Journal* article on the CD phenomenon heaped praise on Sony for its "perseverance." Sony had been so successful in popularizing the CD, according to the *Journal,* that "most Americans believe that Sony—rather than Phillips—invented the technology."

"Sony Battles Back" was the headline on a *Fortune* magazine article, which acknowledged that Sony was "hitting its stride technically" as well as financially. Still, something about the company continued to bother the financial analysts and business journalists. With its constant push for novelty, it was too unpredictable. Unless Sony made even more basic changes, *Fortune* concluded, it was destined to go on "living dangerously from magic show to magic show."

THE KIDS IN THE BLACK PAJAMAS

THE ELDER CITIZENS of the electronics world took a skeptical view, at first, of the American consumer's romance with the VCR. They had seen hot products come and go, and they knew the difference between a passing fling and true love, or so they believed. When VCR sales showed a slight leveling trend around 1979, it seemed to many an esteemed expert that the product had lost its oomph and would soon be supplanted in the public's affections by more sophisticated technologies, such as pay cable and pay per view TV, laserdisc players, direct broadcast satellite systems, or some as-yet-unperfected means of transmitting audiovisual matter over the telephone lines at the individual customer's request. The VCR, compared to all these imminent wonders, was an infernally complicated piece of machinery with an unbecoming number of moving parts. To watch movies on it, a customer had to make two trips to the store, toting a box home and back again. To keep him supplied, the movie companies had to turn out tens or hundreds of thousands of cassettes of each title, although the demand rarely lasted for more than a few months. To the futurologists and technological consultants, some form of "electronic delivery"—in which programming would be beamed into people's homes without the need to create a permanent copy—looked more promising. But the public, as it so often does, forgot to consult the experts.

Americans bought some eleven million VCRs in 1985, or more than thirty thousand on an average day. By the end of the year there was one (or more) in nearly thirty percent of the nation's homes.

> While men of the year and women of the year are being named hither and thither, mostly thither [Tom Shales wrote in the *Washington Post*], someone ought to give a nod to the Thing of the Year: the videocassette, which in the past twelve months has had a tremendous effect on American television viewing and American family life. We have gone from being a television nation to being a video nation. . . . By 1955, you felt naked if you didn't own a TV set. By 1965, you felt a tad underdressed if you hadn't gone to color. In 1975, it began feeling a little nippy if you didn't have cable TV. And 1985 was the year you felt positively indecent unless you had a VCR.

Only a few years earlier Jack Valenti, noting the ease of taping movies off pay cable TV, had predicted a decline in the sale of prerecorded cassettes as the pay cable business grew. Instead, pay cable began to founder just as home video took off, and a curious and unexpected form of synergy arose between the two technologies. "Pre-recorded video sales dip and then soar in cities which have just been cabled," *Variety* reported in January 1985. ". . . Studies show customers buy a VCR when they get cable with the intent of recording films off the wires. Instead, they start renting films at the local vidstore. After a while, they realize they've seen all the films coming over the paycable web in an easier, more convenient way. Out goes the paycable service." *Variety* quoted "one exec" who, "noting the VC in VCR, compared the situation to America's experience in Vietnam, with HBO playing the role of the high-tech juggernaut, and the local vidretailer wearing the black pajamas. Free of the cost of putting up satellites or building billion-dollar cable systems, the small guys will win, said the exec."

"Cable, schmable!" was how George Atkinson summed up the situation. "With cable you're still a passive patsy to somebody else's programming dictates. With HBO and a lot of these other cable guys, the bigger they get, they're really becoming fourth and fifth networks because they have to satisfy most of the people most of the time. The video store lets you choose what you want when you want and even *how* you want. Let's say it's a sports event. You can replay it. You can say, 'Move that back. Let's see it in slow motion.' You're in total control. Choice, time frame—what freedom! *You* tell that television set what to do. You're in an active state versus a passive state. As people find this out, they decide they would perhaps rather spend that thirty-three dollars a month more wisely, go into a store, and, on impulse, depending on their mood, find something—be it esoterica, foreign films, serious movies, classic movies. You can dip into the past, see. It's a ritual and a fun thing to go into a video store. It's not like going to the shoe store to buy shoes. People come in with a positive frame of mind. Kids drag their parents in. I see kids saying, 'Mommy, mommy, I want this one!' And their mommies are saying, 'Let's rent a Fred Astaire.' What we have is the most superior delivery system known to man. It cuts across the generations. It's here to stay. It's not a fad."

Gradually the cable magnates began to shelve their expansion plans and worry about their jobs. A similar shift of mind-set could be observed in the executive suites of the satellite TV moguls, who had been given grand titles, generous expense accounts, and panoramic views of the rooftops of Manhattan to help them launch high-tech schemes for bouncing signals into homes and apartments without the aid of the networks, the local stations, the cable companies, or any other preexisting element of the TV industry. It seemed for a time as if every company that could even remotely be described as being in the "communication business" had some such plan in the works. Then it seemed as if they all had decided to file for bankruptcy at once.

By 1983 the average wholesale price of a VCR was down to $470. Two years later it was $353. By 1986, with the arrival of Korean and Taiwanese models and the discounting of the less popular Japanese models, a consumer could buy a new VCR for $200—not much more than the cost of a typical repair. The disposable VCR was just around the corner. In its first few years on the market the VCR had been a plaything of the rich. Now, according to a study by the consulting firm of Wilkofsky Gruen Associates, some twenty-one percent of families earning less than $20,000 a year had a VCR in the home, while six percent of them had two or more. "In a couple of years," George Atkinson calmly predicted, "it's going to be as commonplace as a bloody toaster."

The video software business was also bouncing along, although it wasn't everybody who could handle the ride. By the mid-eighties the independent video dealer was worrying about regional chains and franchise operations that could rotate inventory, spread advertising costs, and deal directly with the home video companies instead of going through middlemen. Retailers who had started out as small-timers a few years back—people like Frank Barnako in Washington, Linda Lauer in Phoenix, Arthur Morowitz in New York City, Jack Messer in Cincinnati, and Weston Nishimura in Seattle—had built up small regional empires. But they faced stiff competition from bookstores, record stores, department stores, five-and-dime stores, supermarkets, and other interlopers. Tae Kwon, a Korean immigrant who had gone into the dry-cleaning business in Greenwich Village, couldn't help noticing that home video customers, like his own, went back and forth between home and store every few days, so he bought some cassettes, put them in his window, and launched the first video dry cleaners. When two Arkansas supermarket chains began stocking videocassettes, rental prices in the Little Rock area fell to two dollars a night and less—a level at which a number of independent dealers in the area found they could not compete and promptly shut down. In the summer of 1984 the U-Haul Company, blessed with extra floor space as well as idle trucks, put six hundred of its prime outlets into the videocassette business. In the spring of 1986 the Southland Corporation instantly became one of the biggest video retailers in the country with the announcement that it would be renting cassettes at two thousand of its 7-Eleven stores.

In the nine years since George Atkinson had started it all, the inventory of the typical video store had swelled into the thousands of cassettes, and the number of stores had passed the twenty thousand mark (which made them about as plentiful as movie theaters). In 1986 a retailer who wasn't complaining about the glut of stores in his neighborhood—twelve, for example, in a twenty-block section of Los Angeles—was probably complaining about the glut of titles being foisted on him by the home video companies and the need to pump all his profits back into the purchase of cassettes just in order to keep current. With more than fifteen thousand commercial titles on the market, "Enough already!" became the rallying cry of the industry. In the words of Bob Klingensmith, an executive of Paramount Home Video, there was "no more room for *The Man with the Brain in Each Foot.*" But restraint was more easily preached than practiced. If a walk down the aisle of a well-stocked video store took one past *The Spook Who Sat by the Door, Hell River, Hawk of the Caribbean, Stateline Motel,* and *Poor White Trash II,* it was because somebody somewhere expected to make a profit on every title.

A few years earlier it had been common for studio executives to dismiss home video as a marginal business. By the summer of 1986 Paramount had sold 1.2 million cassettes of *Beverly Hills Cop* at a wholesale price in the neighborhood of $20, which added up to gross earnings for the studio and its home video subsidiary of about $24 million. *Beverly Hills Cop* was priced at $29.95 retail, to stimulate the sell-through market. Unlike Paramount, most of the other home video companies continued to price new titles at $79.95 or so, a level at which rentals, inevitably, constituted the overwhelming majority of the business. But rental had become an extremely profitable affair for everyone concerned. Although few consumers were willing to pay out $79.95 to own a copy of *Rambo,* Thorn-EMI-HBO Video sold 427,000 copies to video dealers at a wholesale price of about $52, which translated into about $22 million in earnings for Thorn-EMI-HBO. And when the rental transactions on a particular title began to ebb, the movie companies discovered that they could, after a brief moratorium, reissue the same movie at a lower price—a sell-through price—and make some nice extra money. In this way Hollywood managed to achieve by marketing strategy much of what it had been after in its unsuccessful quest for the legal authority to establish a two-tiered market, divided between sales and rentals.

"This industry is prone to obscene profits," Leonard White, the president of CBS/Fox Video, remarked, without apparent embarrassment, one day in March 1986, when he had just returned to his company's headquarters in New York from a wholesalers' conference in

Acapulco. CBS/Fox Video had made a profit of $60 million the previous year on gross earnings of $250 million dollars—not counting an undisclosed but certainly large sum of money paid to the studio for the use of its movies. When Twentieth Century-Fox needed a $75 million loan that year, it borrowed the money through CBS/Fox Video rather than go to a bank in its own right. The child had become more creditworthy than the parent.

The CBS/Fox profit picture emerged from documents which Fox's new owner, Rupert Murdoch, had been obliged to submit to the Securities and Exchange Commission as part of a stock offering. As a rule, the home video companies guarded their finances closely, in order, perhaps, to avoid giving stars, directors, and producers ammunition to use against the studios when it came time to talk money. But videocassettes were a large part of why Wall Street, in the early 1980s, began to regard Hollywood with new respect and why Fox, MGM, UA, and Embassy figured in a wave of takeovers and mergers.

Sometimes the home video market was kinder to a movie than the theatrical market had been. *The Cotton Club* spent months on *Billboard* magazine's list of hottest-renting cassettes after doing abysmal business at the box office. *Heaven's Gate* earned twice as much money in its video afterlife as in two separate theatrical releases. Movies that people would not pay five dollars to see in a theater they were happy to rent for two or three dollars. As one customer explained to *Time* magazine, "If it's lousy, I just do a fast forward and say goodbye." Science fiction and horror movies frequently fared better in video stores than in theaters. So did the occasional "family movie." A G rating had become such a curse at the box office that producers had been known to throw in an obscenity here and there in hopes of descending a notch or two on the industry's official scale of good taste. But innocence turned out to be something of an asset at the video store as parents discovered the VCR's potential as a child care tool—or had it called to their attention by their children. If the price was right, kid vid, as the genre became known, had an extra appeal for producers and retailers: It sold as well as rented. Disney and Warner discovered a lively demand for low-priced cassettes constructed out of thirty- and forty-year-old cartoons starring the likes of Goofy, Daisy, Pluto, Daffy, Porky, and Bugs. After another change of management at Disney the company decided, in 1985, to abandon its long-standing policy of keeping a select list of children's movies out of circulation between their infrequent theatrical rereleases. When *Pinocchio* appeared on cassette that summer at $79.95, Disney sold 150,000 copies, and when the price was lowered to $29.95, another 250,000 copies were sold.

Kid vid was big business—big enough to attract people without any obvious commitment to wholesomeness for wholesomeness's sake. Noel Bloom, the owner of a company called Caballero Control, which had been built on the allure of such titles as *Talk Dirty to Me,* channeled some of his profits into the creation of a second company, Family Home Entertainment, which trafficked in *Daddy Gumby's Holiday special, G.I. Joe,* and the like. ("We never talk about the two companies in the same breath," a spokeswoman declared.)

Could the machine that made all this possible have *harmed* the movie industry? The passage of a few years had certainly called some of Hollywood's direr warnings—about "entertainment deserts" and such—into question. Home taping and prerecorded cassettes, it became apparent, expanded the audience for movies. Like the discount fares offered by the airlines, they brought in new customers—people who, on a given night, might not otherwise have chosen to watch a movie at all—in numbers that more than offset those who might have gone to a movie theater and paid the full fare. And people who developed a fondness for watching movies at home did not, by and large, stop going to movie theaters. "We've discovered that heavy moviegoers are also heavy cassette buyers," Mel Harris, the president of Paramount Home Video, observed. "It's like getting hooked on chocolate. The more you get, the more you want."

"The history of technology, perhaps more than any other kind of history, is full of premature obituaries," Daniel Boorstin, the librarian of Congress, observed during a talk to the congressional copyright symposium in Florida, just after the *Universal* v. *Sony* decision. "We are prone, especially in this fast-moving country," Boorstin said, "to what I call the displacive fallacy—to believe that every new technology displaces the old technology; that television will replace radio, that electronic news will displace print journalism, that the automobile will displace the human foot, and that television will displace the book. But each of these new technologies has simply given a new role to the earlier technologies. The development of technology is not displacive—it is cumulative." The VCR's impact on the movie and TV businesses seemed, on the whole, consistent with Boorstin's analysis. But while VCRs did not render movie theaters or commercially sponsored TV obsolete, they forced those outlets to adjust to a new and less privileged place in the communications scheme of things. The theatrical movie business, for example, focused more on teenagers and young adults. "You'll always go to the movies if you're between the ages of twelve and twenty-one and you want to make out," Tom Shales pointed out, and he added, "I think the VCR *has* harmed movies, but the real harm is to their quality because they've become even more dependent on a core audience of kids and adolescents and very young adults. The filmmakers have sort of forgotten about appealing to anybody over that age."

The rise of home video was blamed, along with other factors, for a decline in the major networks' share of TV viewership, from ninety percent in 1980 to seventy-five percent in 1984 (when the trend seemed to start leveling out). As home video, cable, pay cable, and Public TV introduced audiences to the notion of the uninterrupted program and broadened the range of material available for viewing, people seemed to grow less content with what the networks gave them. And despite all the protestations of the Home Recorders, TV viewers no longer seemed as tolerant of advertising as they had been. At the end of 1984 a Nielsen survey showed that sixty percent of VCR owners, by their own account, "frequently" or "usually" fast-forwarded through commercials.

A VCR allowed a viewer to get more value from his TV set. "Parents can control what their children see," Tom Shales observed. "They don't just turn it on. They say, 'Here, watch this!' Studies have shown that people who add onto their TV set with VCRs and such end up watching more TV, not less. What the studies don't show is whether they end up watching television more discriminately, more intensely, and more selectively—and I have a feeling that they do." But in the traditional economic structure of television, there was no obvious way for advertisers, broadcasters, or producers to turn this higher-quality service into more revenue. The more discriminating, the more intense, the more selective a TV viewer, the less likely he was to sit compliantly through a commercial he had not asked to see.

By 1985 it took fifteen spots to make the same impression on an audience as ten spots under the old order, according to an executive of the advertising firm of J. Walter Thompson. Fear of "zapping" became an obsession. Burger King produced a series of "speedproof" ads designed, it was said, to get the message across even when viewed at an accelerated pace by an impatient VCR user. NBC went this trick one better with its broadcast of the World Cup soccer matches; it made the program and the commercial inseparable, by parking ads in a corner of the screen while the game continued around them. Other sponsors turned their commercials into epic productions lush with special effects and exotic locations in the style of Steven Spielberg and George Lucas.

The audience ratings firms, responding to the anxiety of advertisers, began studying viewing habits more meticulously than ever. No longer content to know that a certain fraction of viewers was tuned to a certain program at nine o'clock, they sought to determine how many viewers stuck with the program to the end and how many stayed put for the ads—an issue the advertisers hadn't seemed all that anxious to probe in the pre-VCR years. With "scanning wands" that could be waved over the product identification codes in a load of supermarket purchases, it even became possible to answer the ultimate advertiser's question, Did the viewer of Commercial A go out and buy Product A? And in the works, according to a story in *Newsweek,* were plans "bizarre enough to boggle James Bond: ultrasonic devices installed on the set to detect heart rates; tiny, signal-transmitting electronic 'bugs' implanted in a viewer's navel; even living-room couches wired with infra-red sensors to record the varying temperatures of different derrieres."

Together, the technology of television and the idea of commercial sponsorship had produced a close-knit industry in which a small circle of companies routinely did business with each other. The VCR was one of a number of developments that came along in the seventies and eighties to make things less cozy and predictable. Whether that added up to *harm* or not was a matter of definition. That it put pressure on the three networks and their subcontractors—companies like MCA, which had grown large and prosperous under the existing arrangements—was undeniable.

And so was the fact that half-inch video had become the preferred medium of movie pirates the world over. In the Middle East, South and Central America, the Philippines, Indonesia, Malaysia, and Singapore, pirates dominated the home video business, and they weren't very secretive about it. In the fall of 1985 *Variety* estimated the number of illegal video stores in Cyprus at 600—approximately 1 for every 350 households.

> The world of piracy in the Middle East is not without its own bizarre twists [the *Variety* article went on to observe]. One recent example is the circulation of "Rambo" with French and Arabic subtitles that use a different storyline. Set in the Philippines in 1943, the subtitled version has Rambo returning, after escaping two years earlier, to rescue World War II POW's still held by the Japanese. All dates are changed accordingly, despite action scenes with rocket-firing helicopters and other high-tech weaponry. When a U.S. soldier tells Rambo on the English soundtrack: "You made a hell of a reputation in 'Nam," the subtitle read, "You made quite a name for yourself in Guadalcanal . . ." "It's obviously been done for political reasons," said a diplomat who viewed the pirated tape, noting the subtitles delete all references to Vietnam and the Soviet Union.

In Thailand and parts of Africa, law and local tradition allowed public showings of a movie without the copyright owner's approval, so concessionaires could truck around with a VCR, a TV set, and (where necessary) an electric generator, setting up instant movie theaters in villages that had no other access to film or video; and the movie industry went uncompensated, except for the price of the prerecorded cassette, if it happened to be a legal one. In India people gathered to watch movies in cafés, restaurants, and storefront video theaters. Brazil and other financially pressed countries passed laws requiring a hefty quota of home video material to be produced domestically. But instead of keeping American movies out, as intended, these rules served mostly to stimulate the growth of a black market. Restrictions on the flow of money abroad for nonessential purposes had the same effect.

As time passed, the movie industry had some success in getting foreign governments to wean their peoples off the illegal product and onto the legal. But like the United States in its youth, the nations of the third world did not always see copyright law as beneficial to their development. Jamaica, for example, had taken the satellite signals of Home Box Office and other American pay TV services and sent them out on its national television system, free of charge. The Jamaican government expressed a willingness to pay but balked at the MPAA's insistence that it edit out movies that had yet to play in Jamaican theaters. "They want movies that American television viewers have never seen!" Jack Valenti protested. "Because it was shown in their homes by the Jamaica Broadcasting System for a couple of years before we tried to stop them, they begin to believe that it is their just due."

A similar problem of attitude confronted the MPAA's antipiracy efforts at home. In rural parts of the United States the purveyors of satellite dishes had equipped hundreds of thousands of Americans to receive cable TV signals without charge, and the buyers and sellers of those systems were on the political rampage against scrambling plans which would make them pay for what they had been getting free. "I sat down the other day with a man who represents these backyard satellite dishes," Valenti said, "and he said to me, 'That picture is coming through the air. Our people have a right to it.' And indeed there are bills in Congress now to stop HBO from scrambling its signal. It's incredible! There is this notion that's gaining more and more credence, that anything that's up in the heavens is fair game and I have a right to waylay it and bring it into my home. Because they've been receiving this free, if you say, 'Wait a minute, you have to pay for it,' they act as if you have committed rape on their virginal daughters."

The MPAA was also fighting guerrilla raids on the conceptual border between a private and a public performance. Robert Zeny, a video dealer in Erie, Pennsylvania, decided to set up "viewing rooms"—each containing a couch and a TV monitor—for customers to watch rented cassettes in. A lawsuit brought by Columbia, Embassy, Paramount, Fox, Universal, Disney, and Warner closed Zeny's business down and foiled his plans to franchise. But the idea refused to die such an easy death. A few months later the studios were back in court against another viewing room impresario, John Leonard of State College, Pennsylvania, who claimed that *his* store—the Nickelodeon Video Showcase—was legal since the customers rather than the employees ran the VCRs. "The movie is performed to the customer, his family and his social acquaintances," Leonard said. "We are not showing films in public." To the movie industry's great relief, however, a federal judge thought otherwise.

By the mid-eighties MCA was as deeply committed to home video as any of the studios, and its home video division was one of the healthiest units of the company. That fact did not lead Sidney Sheinberg to conclude that he had misjudged the VCR's impact on his industry. "With the benefit of hindsight, I think it has been even more harmful than we thought," Sheinberg said. "The harm is not only in the copying of material, which deprives us of subsequent potential revenues—all the arguments that we made in the litigation—but in the continuing degeneration of the concept of copyright. Whether it's people plucking the HBO signal off the air or not paying for taps on cable systems or whatever, it's caused and fed a deteriorating respect for a basic and constitutionally motivated right. This is a worldwide problem that our industry faces. Now, even the plague has some positive aspects to it. Scholars are always writing treatises on its role in the industrial revolution. But if you said to me, 'Hey, would you like to turn the clock back and undo it all?' it wouldn't take me a fraction of an instant to answer." He paused just longer than a fraction of an instant and added, "But you can't."

MCA, like the other studios, was starting to produce works intended exclusively for release as videocassettes. It had announced a series of feature-length "thriller-terror" tales to be produced for less than a million dollars each and released on a monthly basis. It was also preparing an assortment of "how-to" programs, including a yoga tape entitled *Yoga Moves* and an adult sex education tape, *Love Skills*, which, according to one MCA executive, would include full-frontal nudity and simulated intercourse but be "so clearly respectable, solid, and informational that no one could possible confuse it with pornography." By 1985 the flow of theatrical movies into home video was slowing down, while the output of original programming, mainly in the how-to and kid vid genres, was speeding up; and the studios were doing their best to adjust.

Still, it was not every Hollywood executive who could make the necessary mental leap. And the studios didn't seem to have their hearts in it as much as some of the home video pioneers who moved in from outside the feature film business—people like Stuart Karl, a young Californian who migrated from water beds to home video and wooed Jane Fonda into letting him produce the videocassette of her exercise book, *Jane Fonda's Workout.* In four years the Fonda cassette sold more than a million copies, while setting off a video exercise binge. Soon one could work out with Richard Simmons, Jack LaLanne, Debbie Reynolds, Bubba Smith, Lyle Alzado, Marie Osmond, Jayne Kennedy, Irlene Mandrell, Jake (of Body by Jake), or Raquel Welch. And one could cook with Julia Child, practice tennis shots with John Newcombe, watch birds with Roger Tory Peterson, or file for divorce with Marvin Mitchelson, the pioneer of palimony.

Around the same time that Stuart Karl got together with Jane Fonda, Austin Furst, a former executive of Time Inc., purchased the library of Time-Life Films, a high-toned collection of documentaries and features, and launched his own company, Vestron Video, based in Stamford, Connecticut. Then Vestron made a deal for the rights to *Making Michael Jackson's "Thriller,"* a home video original that proceeded to sell nearly a million copies at thirty dollars each, although it had been turned down by CBS/Fox because no previous music video had sold more than seventy thousand copies. What with one thing and another, Vestron was soon selling more cassettes than most of the Hollywood studios, and by the summer of 1985 Furst had personally earned eighteen million dollars and held stock valued at nearly half a *billion* dollars. In January 1986 Vestron let the word go forth that it was about to start producing theatrical movies as well as videocassettes.

Home video was a port of entry into the movie business, which became more open and fluid than it had been in decades. After he sold his interest in Magnetic Video to Twentieth Century-Fox, Andre Blay stayed on as president until the summer of 1981. Then things turned unpleasant. "It was a classic example of the bought-out executive who just didn't like the new owners," Blay said later. "I couldn't tolerate their management style, and life's too short. So I said, 'Let's call it quits.' The way they handled people, the way they handled budgeting, the way they handled marketing. The business was a runaway success, and they wanted their hands on it. It got to be corporate politics, so I went to the south of France for a couple of months and studied French and drank a lot of wine and espresso."

When he had had his fill, he set up a new company, the Andre Blay Corporation, with a tiny staff working out of offices in Northville, Michigan. Within a year and a half Blay had made home video deals with Lew Grade, Joseph E. Levine, and the Rank organization, and he had put out a successful music video starring Elton John as well as an "adult video magazine" in partnership with *Lui* magazine and the French edition of *Playboy.* Then Embassy Communications bought him out, merging his company into a new entity known as Embassy Home Entertainment, and Blay moved to Los Angeles as its chairman, a job in which he found himself reading ten scripts a week and doling out millions of dollars at a shot for the home video rights to movies which, in many cases, would not be made without his approval. When Coca-Cola bought Embassy in 1985, Blay and a few of his fellow executives banded together in an effort to buy Embassy Home Entertainment for "a little bit more" than eighty million dollars, and Blay said he planned to run the venture "more and more like a movie company."

George Atkinson was another pioneer who surrendered control of his business and had to start fresh. By 1983 the Video Station had five hundred affiliated stores in addition to the five it owned outright, and the company was making fifteen million dollars a year, mostly as a wholesaler of videocassettes. But when questions were raised about the company's financial management, the board of directors appointed a three-man executive committee to set things right, and when the committee refused to approve some of Atkinson's expansion plans, he set off on his own. "I got a taste for manufacturing, because I figured that those were the sweetest margins besides retail," he said. He began with an extremely modest library, consisting of a few old movies whose copyrights had lapsed through the indifference or inadvertence of their proprietors, and a few foreign movies whose rights were available on the cheap. His titles included *The Little Princess* with Shirley Temple, *The Big Trees* with Kirk Douglas, and the Soviet version of *War and Peace*—a two-cassette package encased in simulated leather to give it the look of a collector's edition of the book.

Manufacturing videocassettes, Atkinson discovered, was a business he could carry on largely over the telephone. "I pick up the phone and I talk to the producer, and we contract the rights to a picture," he explained. "Then I need to get a print—a good print. We make a master and we give it to a lab—say Bell and Howell—and tell them, 'Strike so many half-inch pieces for us.' They do the dirty work. Then we have to make the jacket covers. I like to get involved in designing the thing. I have my artists. I say, 'Lay it out this way,' and I proof it and give my okay. Then I have my network of distributors—my pipeline—and another phone call or a mailout lets them know what's coming. They do their ordering via the phone, and we either drop-ship out of the lab or do a little minor shipping at our end. We're kind of traffic-copping the whole thing, see. Our overhead is low, and we're making wonderful margins. Right now this is a three-man operation, but we can do millions of dollars' worth of business. Okay, it's not *Raiders of the Lost Ark*, but for every locomotive there are two hundred boxcars."

Atkinson explained these matters during a break in the fourth annual convention of the Video Software Dealers Association, at the Washington Sheraton Hotel in August 1985. With all its vital statistics rising, the home video industry treated the occasion as a coming-out party. Vestron Video announced the creation of an Academy of Home Video Arts and Sciences, which would issue awards for the best in made-for-home-video programming. Six thousand dealers attended the convention, and so did Lena Horne (with the home video version of her one-woman Broadway show), Jane Fonda (with her fifth exercise cassette, the *New Workout*), and Joan Collins (with a series of old Hollywood tearjerkers glossily repackaged as *The Joan Collins Home Video Collection*).

Here was a Washington gathering marred by no talk of federal deficits, no cries of alarm about the balance of trade and the decline of

American manufacturing, and, for sure, no tirades against the Japanese, the people who had made it all possible. If you were in the video business and you weren't making plenty of money, you had no one to blame but yourself. If you weren't in the business, well, now was the time to get in, and the VSDA convention was the place. Among the accessory products and services being dispensed there were a gag music video entitled *The Pet Rock Video,* with a dimple-faced slab of granite as the star; an assortment of computer programs for managing a video store's accounts; the security expertise of a certified ex-con ("Hi, my name is Mike McNasty, and I used to be a thief"); several brands of videocassette vending machines; magazines called the *Video Store,* the *Video Software Dealer, Video Times,* and *American Video Monthly;* and a series of "Giftvideo Greetings," whose producer described them as "*the* gifts for the eighties—an imaginative and new way to celebrate life's memorable occasions."

In the how-to genre, *Esquire* magazine unveiled a line of *Esquire Success* tapes whose topics included "Persuasive Speaking," "The Short-Order gourmet," "The Wine Advisor," and "Career Strategies 1" and "Career Strategies 2," which promised to teach "young achievers" all about "developing managerial skills," "establishing a power base," and "knowing when to challenge the system." Each tape came with a "video essay" by Dick Cavett affixed. Young professionals are "brand-name oriented, and they have no alternative in the video field," Philip Moffitt, the president and editor in chief of *Esquire,* explained. "No one, and I mean absolutely no one, has established a brand-name label for this particular audience."

Arlene Winnick, a fledgling video magnate from Long Island, came to the convention with a lineup of videocassette games which ranged from Video Bingo to Videotrivia—"a perfect rental item," she explained, since "people come into the store and say, 'What will we do tonight? Oh, well, let's play trivia.' " Winnick confided that she was not impressed with much of the material being offered at the surrounding booths. "I couldn't believe it when I heard about it!" she said of the *Pet Rock Video.* As for the *Esquire Success* tapes, "They'll get a few yuppie mailing lists and they'll do well" was her cool assessment. "Everyone has their own little niche and there's plenty of room in this business," she said. "As long as I get my niche, I don't care."

The movie studios had divided up the honor of feeding the dealers, with Walt Disney providing a lunch that featured a Radio City Music Hall-style song-and-dance show and Paramount Pictures hosting a "Star-Spangled Picnic" under a vast tent on the grounds of the Washington Monument, followed by a concert at Constitution Hall. Even Jack Valenti and Charlton Heston put in appearances—and why should they not? Valenti was about to become a home video star himself with the release of a package of how-to cassettes based on his book *Speak Up with Confidence.* (The complete set was priced, with the rental market in mind, at $1,499.) The gross revenues from the sale and rental of videocassettes—a projected three and a half billion dollars by 1985's end—were closing in fast on the gross revenues of the theatrical movie business. Or as George Atkinson observed, "We're catching the big daddy." And that's at twenty-five percent penetration. From there you can just extrapolate!"

SOURCES

ANTIROYALTY CAMPAIGN IN WASHINGTON

John Adams, William Baker, Michael Blevins, Jack Bonner, Ronald Brown, Robert Bruce, Nancy Buc, Joseph Cohen, Marlow cook, Nina Cornell, Jeffrey Cunard, Charles Ferris, Carol Tucker Forman, Hall Northcott, David Rubenstein, Alan Schlosser, Gary Shapiro, Robert Schwartz, William Tanaka, Dennis Thelen, Jack Wayman.

CARTRIVISION

I. Walton Bader, Lawrence Hilford, Jeffrey Reiss, Frank Stanton. Books and periodicals: *Television Digest, Wall Street Journal.*

CONGRESS AND LOBBYING

Michael Apitoff, Ed Baxter, Howard Berman, Dan Buck, Dennis DeConcini, Elizabeth Drew, Don Edwards, Roberta Haberle, Randy Huwa, Vince Lovoi, Robert Kastenmeier, Peter Kostmayer, Deborah Levey, Charles Mathias, Thomas Mooney, Thomas Olson, Ralph Oman, Andrea Pamfilis, Michael Remington, Michael Synar.

Books and periodicals: *New York Times, Washington Post, Variety, Video Week.*

COPYRIGHT AND PATENT LAW

Daniel Boorstin, Peter Jaszi, David Ladd, David Lange, Kate McKay, Alan Latman, Melville Nimmer, Linda Garcia Roberts, Stanley Rothenberg, Leon Seltzer.

Books and articles: James J. Barnes, *Authors, Publishers and Politicians: The Quest for an Anglo-American Copyright Agreement* (Ohio State University Press, 1974); Albert J. Clark, *The Movement for International Copyright in Nineteenth Century America* (Greenwood, 1960); Benjamin Kaplan, *An Unhurried View of Copyright* (Columbia University Press, 1967); David Lange, "Recognizing the Public Domain," *Law and Contemporary Problems* (Fall 1981); Lyman Ray Patterson, *Copyright in Historical Perspective* (Vanderbilt University Press, 1968); Arnold Plant, *The New Commerce in Ideas and Intellectual Property* (University of London/Athlone Press, 1953); Dennis Welland, *Mark Twain in England* (Chatto and Windus, London, 1978).

EVR

Robert Brockway, Morton Fink, Leon Knize, Richard Rosenbloom, Harry Smith.

Books: Peter C. Goldmark with Lee Edson, *Maverick Inventor* (1973); William S. Paley, *As It Happened* (Doubleday, 1979).

HISTORY AND TECHNOLOGY OF VTRS AND MAGNETIC RECORDING

Albert Abramson, George Brown, E. Stanley Busby, Harold Clark, Bruce Follmer, Charles Ginsburg, Peter Hammar, David Jensen, David Lachenbruch, Ralph Mossino, Tom Owen, Joseph Roizen, Mark Sanders, Mark Schubin, Ken Winslow.

Books and articles: Albert Abramson, "A Short History of Television Recording," included in *Technical Development of Television,* an anthology edited by George Shiers (Arno Press, 1977); Charles P. Ginsburg, "The Birth of Videotape Recording" (a paper delivered before the Society of Motion Picture and Television Engineers, 1957); Ginsburg, "The Horse or the Cowboy: Getting Television on Tape," a lecture to the Royal Television Society, November 5, 1981 (reprinted in its journal, *Television* [November/December 1981]); Harold Lindsay, "Magnetic Recording," *db: The Sound Engineering Magazine* (December 1977 and January 1978 issues); John T. Mullin, "Creating the Craft of Tape Recording," *High Fidelity* (April 1976); Shigeo Shima, "The Evolution of Consumer VTRs," *IEEE Transactions on Consumer Electronics,* (May 1984); Yuma Shiraishi, "History of Home Videotape Recorder Development," *SMPTE Journal* (December 1985); Hiroshi Sugaya, "Home VTR Made Possible," *Business Japan* (June 1980).

HOME VIDEO BUSINESS

Harry and Connie Allen, Frank Amato, Alan Anton, George Atkinson, Frank Barnako, Katrine Barth, Andre Blay, William Brooks, Doug Burns, Joseph Cohen, Rocco LaCapria, Morton Fink, Mel Harris, Lawrence Hilford, Bill Holland, James Jimirro, Leon Knize, Linda Lauer, Jack Messer, Kathy Montagna, Arthur Morowitz, Leo Murray, Weston Nishimura, Steve Roberts. Bill Silverman, Risa Solomon, Russ Solomon, Paul Wagner, Leonard White, Arlene Winnick.

Books and periodicals: *Video 1995* (Wilkofsky Gruen Associates, 1985), *Variety, Video Week.*

JAPAN AND JAPANESE BUSINESS

James Abegglen, Herman Bertsch, Herbert Cochran, Bill Finan, David Halberstam, Hideo Hirayama, Tetsuo Hirayama, Tomosaburo Iwayama, Bernard Krisher, Richard Samuels, Jim Shinn.

Books: Robert C. Christopher, *The Japanese Mind* (Simon and Schuster, 1983); Rodney Clark,

The Japanese Company (Yale University Press, 1979); Chalmers Johnson, ed. *MITI and the Japanese Miracle* (Stanford University Press, 1982); O-Young Lee, *Smaller Is Better: Japan's Mastery of the Miniature* (Tokyo: Kodansha International, 1984); Charles J. McMillan, *The Japanese Industrial System* (Walter de Gruyter, 1984).

JVC

Toshihiro Kikuchi, Shuji Kurita, Masayuki Murakami, Makoto Nakamura, Yuma Shiraishi, David Walton, Junko Yoshida.

MATSUSHITA

Yancy Fukagawa, Yasuyuki Hirai, Tsuzo Murase, Masao Nakata, Tak Shiraki, Ken Shimba, John Sprague, Masaharu Takahara, Ian Watanabe.

Books and periodicals: Jeffrey Cruikshank, "Matsushita," *The Harvard Business School Bulletin* (1983); Rowland Gould, *The Matsushita Phenomenon* (Tokyo: Diamond, 1970); Konosuke Matsushita, *Not for Bread Alone: A Business Ethos, A Management Ethic* (Tokyo: PHP Institute, 1984), and *Thoughts on Man* (Tokyo: PHP Institute, 1981)

MOVIE INDUSTRY—HISTORY

James Bouras, Peter Jaszi.

Books and articles: Gordon Hendricks, "The History of the Kinetoscope"; Robert Merritt, "Nickelodeon Theaters 1905–1914)"; Robert Anderson, "The Motion Picture Patents Company—A Reevaluation"; Douglas Gomery, "U.S. Film Exhibition—The Formation of a Big Business"; and Mae D. Huettig, "Economic Control of the Motion Picture Industry." The foregoing are contained in the anthology *The American Film Industry,* edited and annotated by Tino Balio (University of Wisconsin Press, 1976 and 1985).

PROROYALTY CAMPAIGN IN WASHINGTON

Fritz Attaway, Michael Berman, Timothy Boggs, James Bouras, James Corman, Barbara Dixon, Ewing Layhew, Bruce Lehman, William Patry, Thomas Railsback, Romano Romani, Rob Smith, Dale Snape, Robert Strauss, Jack Valenti, Joseph Waz, Anne Wexler.

RCA

George Brown, Howard Enders, William Hittinger, Frank McCann.

Books and periodicals: George Brown, *And Part of Which I Was: Recollections of a Research Engineer* (Angus Cupar, 1983); David Sarnoff, "The Fabulous Future," *Fortune* (January 1955).

SONY

Morton Fink, Ira Gomberg, Akinao Horiuchi, Masaru Ibuka, Miki Iwaya, Nobutoshi Kihara, Haruyuki Machida, Teruo Masaki, Akio Morita, Masahiko Morizono, Yasuzo Nakagawa, Masa Namiki, Jiro Ohbu, Richard O'Brion, Harvey Schein, Roger Stone, Kenji Tamiya, Koichi Tsunoda, Fred Wahlstrom, Tom Sugiyama, Kiyoshi Yamakawa.

Books: Nick Lyons, *The Sony Vision* (Crown, 1976); Akio Monita, *Made in Japan* (Dutton, 1986); Yasuzo Nakagawa, *The Development of Semiconductors in Japan* (Tokyo: Diamond, 1981), and *The History of Magnetic Recording in Japan* (Tokyo: Diamond, 1984); Kazuto Mimura, *Sony's Counterattack* (Tokyo: Yamate Shobo, 1984); and *Genryu: Sony Challenges, 1946–1968 (Sony Corporation, 1986).*

UNIVERSAL (MCA)

Sally Bartell, Robert Hadl, Lee Isgur, Sidney Sheinberg, Herb Steinberg.

Books: Dan Moldea, *Dark Victory:* Ronald Reagan, MCA and The Mob (Viking, 1986).

Periodicals: The Los Angeles Times, Variety, etc.

UNIVERSAL V. SONY

Fred Barbash, Dean Dunlavey, Ambrose Doskow, Max Freund, Linda Greenhouse, Toni House, Fred Kirby, Stephen Kroft, Robert Loewen, Jim Mann, Peter Nolan, Fred Rogers, Joel Sternman.

Periodicals: *The American Lawyer, Los Angeles Times, Variety, Washington Post.*

INDEX

Abegglen, James, 141, 142
AC-bias recording, 41, 45
Adams, John, 218–20
advertisers, television, 230, 248, 320, 321
Akai, 238
Akin, Gump, Strauss, Hauer, and Feld, 211–12, 229
Albeck, Andy, 207
Alcott, Louisa May, 126
Allen, Harry, 185–86
Allgemeine Elektricitaets Gesellschaft, 42
Allied Artists, 178
Amato, Frank, 184, 185
American Broadcasting Company (ABC), 57
American Film Institute, 275
American Graffiti (film), 27
American Lawyer, 243
American Telephone and Telegraph (AT&T), 46, 47
Ampex Corporation, 57, 70
  collaboration with Sony, 63–66
  home video, 80
  and Japanese companies, 61–63, 66–67
  marketing methods, 69–70
  suit against Cartrivision, 86
  suit against Sony, 68–69, 100
  videorecording, 58–59, 61
Anderson, Charles, 58–59
Anton, Alan, 296–97
Art of Japanese Management, The (Pascale and Athos), 140
ASCAP, 205
Ashley, Ted, 188, 189–90, 207
Athos, Anthony C., 140
Atkinson, George, 175–80, 194, 195, 291, 313–14, 315, 325–26, 328
Attaway, Fritz, 216
audience ratings, 320
audiocassettes, 70, 214
audio rental bill, 301
AutoVision, 92
Avco, 74, 82, 85, 86
Avco Embassy, 174, 207
  see also Embassy Communications
azimuth recording, 90–91

Babycorder, 45
Baker, Howard, 215
Baker, William, 218–19, 223–24

Barnako, Frank, Jr., 180–81, 182, 184, 196, 197–98, 291, 296, 315
Baumgarten, Jon, 214
Begelman, David, 207, 286
Bell & Howell, 64, 65, 74
Bell Laboratories, 46
Benton, Cheryl, 199
Bentson, Lloyd, 258, 260
Berman, Howard, 255
Berman, Michael, 300
Betamax:
advertising campaigns, 21–22, 97, 102–3, 136, 306
auto-changer for, 166
capacity, 147, 153, 160–61, 166
compatibility, 94–95, 146, 147, 150–55, 158–59, 166, 304, 307–8
copyright disclaimer, 35, 242
development of, 91–95
floor-model, 21–22
"high-end" strategy, 307
marketing, 95–97
market share, 165, 167, 304, 306
price, 165
reviews, 166–67
SL-6300, 95–96
SL-7300 deck, 95
SuperBeta, 308
surveys of use of, 107–8
Valenti on, 238
Betamovie, 167
BetaScan, 167
Beverly Hills Cop (movie), 316
Biden, Joseph, 292
Blackmun, Harry, 131, 272, 273
Blay, Andre, 168–69, 172–75, 324–25
Blevins, Michael, 268
BMI, 205
Boggs, Timothy, 250, 257
Boggs, Tommy, 239
Boorstin, Daniel, 318–19
Brennan, William, 269, 272
Breyer, Stephen, 280
Brockway, Robert, 77, 78
Brooks, William H., 183, 184–85
Brown, Bob, 199
Brown, George, 56, 79, 163
Brown, Robert, 165
Brown, Ronald, 239, 240, 249, 250, 290, 296
Bruce, Robert, 231, 249, 264
Brut Productions, 174
Buc, Nancy, 223–24, 249, 289
Burch, Dean, 249, 250
Burger, Warren, 264–65, 269, 272–73
Burger King, 320
Byrd, Robert, 215

Caballero Control, 318
cable TV, 278, 313–14
camcorders, 308–10
Camden, Lord, 123, 124
Camras, Marvin, 57
Canby, William G., 132
Canon, 50
Cartrivision, 80–88
CBS / Fox Video, 195, 202, 317, 324
Cellar, Emmanuel, 254, 281
Cieply, Michael, 305
Clark, Albert J., 125
Cleveland, Harlan, 279
Coca-Cola, 325
Cohen, Joseph, 297, 299
Coleco, 276
Collins, Joan, 326
Columbia Broadcasting System (CBS), 30, 73, 74–81, 195, 210
Columbia Pictures, 82, 182, 207
Comer, Doug, 296–97
Commercial Cutter, 302
commercials, deleted, 248, 302, 320
compact disc (CD), 310–11
computers, 271, 278, 302
Conrad, Anthony, 153
Consolidated Video Systems, 73
Consumer Electronics Show, 178, 197, 221, 236, 310
Consumer Federation of America, 301
Conyers, John, 256
Cook, Marlow, 249, 250

Copyright Act (1909), 128
Copyright Act (1976), 23, 131, 280
copyrightists, 229
  arguments, 229–31
  and campaign contributions, 257
  direct mail campaign, 232–35
  and rental bill, 300
  Supreme Court decision, 269
  and Valenti, 285–86
  see also Motion Picture Association of America; Valenti, Jack
copyright law:
  fair use, 23–24, 103–4
  for films, 128
  and harm, 131–32, 133–34, 230, 248
  history of, 123–29
  international, 125–26
  interpretations of, by lobbyists, 230
  lawyers, 122–23, 129
  and libraries, 129–31
  and money, 255–56
  rentals of videocassettes, 285–88
  revision (1976), 23, 104, 131, 254, 280, 281
  royalty on VCR sales, 205–7, 224–25, 257–59, 261
  and technology, 127–32, 243, 277–83
Copyright Notices, 270
Copyright Royalty Tribunal, 214
Corman, James, 217, 250, 268
Cornell, Nina, 249, 250, 253
Cotton Club, The (movie), 317
Crosby, Bing, 56, 57
Crossley Surveys, 108
Cunard, Jeffrey, 290, 291, 296
Cutler, Lloyd, 249, 250
CV-2000 (Sony VTR), 35, 67–68, 92, 110

Daily Variety, 112, 118, 121
  see also Variety
D'Amato, Alfonse, 204
Davis, John, 106, 215
Davis, Marvin, 194–95
Day, J. Edward, 220, 222, 244, 250, 268
DeConcini, Dennis, 203–4, 206, 213, 218, 219, 224, 260, 288–89, 290, 294, 298, 299
De Laurentiis, Dino, 22, 275
Dickens, Charles, 125
Diller, Barry, 207
DiscoVision, 28, 31, 36, 206
Disney, *see* Walt Disney Productions
Dolby, Ray, 58
Dole, Robert, 257–58, 260, 294, 296, 296–97, 300
Donkey Kong (video game), 275–76
Doubleday and Company, 277
Doyle Dane Bernbach, 21–22, 34, 51, 96
Duncan, John, 203, 206
Dunlavey, Dean, 100–102
  appeal of *Universal v. Sony,* 132–33
  arguments in *Universal v. Sony,* 113
  discovery in *Universal v. Sony,* 109–11
  first oral arguments before Supreme Court, 241–42, 243–44
  reargument before Supreme court, 264–66
  research for *Universal v. Sony,* 103–5, 107
  on Supreme Court, 246

East, William G., 132
Eastwood, Clint, 226
Edison, Thomas, 169, 188–89
Edwards, Don, 215–16, 226, 251
Eisner, Michael, 207
Eizenstat, Stuart, 249, 250
Electronics Industry Association (EIA, U.S.), 219–23
  and rental bill, 288
  on royalty compromise, 261
Electronics Industry Association of Japan (EIAJ), 71, 72, 239–40, 248–49, 296
electronic video recording (EVR), 75–79, 80
Embassy Communications, 325
  see also Avco Embassy
Embassy Home Entertainment, 325
England, copyright in, 123–24
Eslinger, Lewis, 35
Esquire (magazine), 327
Esquire Success (videos), 327
exercise tapes, 324

fair use, 23–24, 103–4
Faith Center Church (Glendale, Calif.), 271
Family Home Entertainment, 318
Federal Communications Commission, 26, 59, 109
Fenster, Ray, 180
Ferguson, Warren, 108–9, 112–21, 133
Ferris, Charles, 222–23, 225, 226, 247, 248, 250, 253, 262, 263
Field Research, 107–8, 116, 120
Filmways, 187
Fink, Morton, 163–64, 187–97
first sale, doctrine of, 179, 290
Fonda, Jane, 324, 326
Forbes (magazine), 305
Foreman, Carol Tucker, 249, 289, 301
Fortune (magazine), 53, 137, 165, 311
Fox, *see* CBS / Fox; Twentieth Century-Fox
France, copyright in, 125
Frank, Barney, 255, 257
Frito-Lay, 248
Furst, Austin, 324

Gallagher, William, 202
gaman, 49
General Electric, 165, 220
Germany, copyright in, 125
Gibson, Dunn, and Crutcher, 100, 101, 274
  see also Dunlavey, Dean
Gillette Company, 117–18
Ginsburg, Charles, 57
Glickman, Dan, 220
Goldmark, Peter, 75–79
Gomberg, Ira, 34, 100, 105, 219, 227, 264, 269, 291
Goodman, Julian, 119
Gortikov, Stanley, 214, 247–48, 261
Gould, Jack, 77
Grade, Lew, 174, 325
  by Home Recording Rights Coalition, 235–37, 240, 290–92, 300
  by MPAA, 232
  by video dealers, 235–37, 240, 290–92, 295–98
Greenspan, Alan, 249, 250
Grier, Robert, 124
Griffiths, Edgar, 160–61
Griffiths, William, 32–33, 34, 117
Gulf and Western, 210–11
Gundy, Phil, 63, 65

Hanaford, Peter, 293
Handycam, 309
Harada, Akira, 161
Harris, Joel Chander, 126
Harris, Mel, 278, 318
Hart, Gary, 286
Harte, Bret, 126
Harvey, James, 207
Hatch, Orrin, 294, 297
Hathaway, William, 250
Hauser, Gustave, 279
Heaven's Gate (movie), 317
helical scanning, 62, 73
Heston, Charlton, 226–27, 233–34, 328
Hilford, Lawrence, 81, 83–84, 85, 88, 202
Hirose, Bunichi, 45
Hirschfield, Alan, 194, 207, 286–87
Hitachi, 47, 72, 156–57, 158–59, 238–39, 308, 309, 310
Hoffman, Alexander, 277–78
HoloTape, 79–80
Home Box Office, 113, 313–14, 322
Home Recording Rights Coalition, 218–26, 229, 235–40, 248–49, 257, 260–61, 288–96, 300
Horne, Lena, 326
Hotten, John Camden, 126
House, Toni, 244, 245, 268
how-to tapes, 324, 327

Ibuka, Masaru, 30, 37–51, 63, 70, 71, 91, 93–94, 147, 149–50
Ideal Toy Corporation, 127
Ikegami, 73
Inai, Takayoshi, 151, 153, 160, 161
InstaVision, 80
intellectual property, *see* copyright law
interlibrary loan, 129–30

International Business Machines (IBM), 52–53
International Tape Association, 163
Isikoff, Michael, 298
Iwata, Kazuo, 150, 151–53, 159

Jamaica Broadcasting System, 322
jamming devices, 119–20
Jankowsky, Joel, 212, 213, 217
Japan:
  acceptance of "voluntary" quotas, 260
  contracts in, 31
  Diet, 238
  industrial development in, 46–47
  Ministry of International Trade and Industry (MITI), *see* MITI
  post-war, 39–40, 42
Japan Broadcasting Corporation (NHK), 40–41, 61
Japan Measuring Instrument Company, 37
  see also Sony
Jaszi, Peter, 205
Jaws (film), 27
Jimirro, James, 192
John Adams Associates, 218–20
JVC (Victor Company of Japan), 143–44
  and Betamax compatibility, 94–95, 143, 146, 150–55, 158
  camcorder, 309
  develops VCR, 145–47, 148–49
  eight-millimeter format, 308
  and standard for VCRs, 72, 73, 89
  videocassette recorder development by, 92–93
  video home system (VHS), 151–53, 159, 162
  videotape recorder (VTR), 62, 143
J. Walter Thompson agency, 320

Karl, Stuart, 324
Kastenmeier, Robert, 104, 212, 216, 226, 253–57, 258, 278–83, 286, 287, 294
Kazen, Abraham, Jr., 104
Kennedy, Edward M., 225
Kerr, James, 86
Kihara, Nobutoshi, 42–43, 45, 62, 67, 71, 72, 73, 89–91, 93, 154–55
Kilkenny, John F., 132, 133, 134
Kindness, Thomas, 255
kinescope recording, 55
King Kong (film), 22, 275–76
Klingensmith, Bob, 316
Knize, Leon, 190–92, 196, 197, 199–200
Kodak, 74, 80, 308
Kostmayer, Peter, 250
Kroft, Stephen:
  appeal of *Universal v. Sony,* 133
  arguments in *Universal v. Sony,* 109, 114–20
  discovery in *Universal v. Sony,* 109–11
  and Doyle Dane letter, 23, 24
  first oral arguments before Supreme Court, 243
  and Griffiths, 33, 117
  reargument before Supreme Court, 266–67
  research for *Universal v. Sony,* 32, 106–8
Kwon, Tae, 315

LaCapria, Rocco, 198–99, 200
Lachenbruch, David, 34, 84
Ladd, David, 131–32, 212
Landis, Ben, 305
Lange, David, 123, 280
laserdisc, 206
Latman, Alan, 130
Lauer, Linda, 182, 200–201, 315
lawyers, copyright, 122–23
Laxalt, John, 293
Laxalt, Michelle, 294
Laxalt, Paul, 294, 299
Leahy, Patrick, 252, 294
Lear, Norman, 109
Leavy, Deborah, 250
Lee, O-Young, 47
Legal Times of Washington, 235
Lehman, Bruce, 249, 257, 286, 300
Leonard, John, 323
Levine, Joseph E., 325
libraries, and copyright, 129–30
"library," home video, 107

Life (magazine), 74
lobbying and lobbyists, 228–29
  appearance of, 251–53
  door openers, 293
  grass-roots, 228–29, 232–37, 240, 290–92, 300
  see also names of specific lobbies and lobbyists
Long, George, 63, 64, 65
Los Angeles Times, 243, 272
Lowe, James, 106–7
Lowell, James Russell, 125
Lui (magazine), 325

Macmillan, Harold, 45
McNair, Robert, 259
Macy's, 85, 142
Made in Japan (Morita), 102
Magnavox, 74, 165
Magnetic Video, 172, 175, 177, 178, 181, 324–25
  see also Blay, Andre
Magnetophons, 42
Making Michael Jackson's "Thriller" (video), 324
Mangano, Michael, 96
Mann, Jim, 243, 272
Marshall, Thurgood, 272
Masaki, Teruo ("Ted"), 237–38, 264
Mathias, Charles, Jr., 212–14, 215, 249, 252, 277–83, 294, 299
Matsuno, Kokichi, 145, 151–53, 158
Matsushita, Konosuke, 137–39, 149, 150–51, 152, 153–54
Matsushita, Masaharu, 94, 150, 160
Matsushita Electric Industrial Company, 137–42
  and Betamax compatibility, 94–95, 143, 146–47, 150–55
  cartridge-format recorder, 92
  competition with Sony, 142–43
  and copyright lobbying effort, 238–39
  divisions, 140–41
  eight-millimeter format, 308, 310
  and JVC, 72
  marketing, 142
  Matsushita Electric Corporation of America (MECA), 142, 161, 220, 223, 261
  OEM deals, 140, 142, 308
  profits, 229
  and RCA, 161–63, 164
  and standard for VCRs, 72, 73, 89
  VCR market share, 224
  and VHS, 157, 159
  videotape recorder (VTR), 62, 146–47
Matsushita Kotobuki, 148, 150, 151, 153
Maverick Inventor (Goldmark), 75
Mazzoli, Romano, 255
MCA, 21, 25–27, 74, 80, 210, 323–24
Meese, Edwin, 293
mergers, 317
Messer, Jack, 181, 182, 299, 315
Metro-Goldwyn-Mayer (MGM), 127, 182, 207
  home video rentals, 202
Metzenbaum, Howard, 294
Mimura, Kazuto, 307
Mini Video Movie, 309
Minolta, 309, 310
Mintz, Levin, Cohn, Ferris, Glovsky, and Popeo, 222
"Mister Rogers' Neighborhood" (TV program), 106, 270, 271
MITI (Ministry of International Trade and Industry, Japan), 47–49, 61–62, 66, 157–58, 260
Mitsubishi, 47, 157, 158, 238–39, 309
Mittleman, Sheldon, 22
Moffitt, Philip, 327
Montgomery Ward, 85, 165
Mooney, Thomas, 256
Moorhead, Carlos, 251, 255
Morita, Akio, 38–39, 99
  and Ampex, 53
  annual shareholders' meetings, 307
  and Betamax compatibility, 147, 148, 150–55
  and Betamax development, 93–94
  and copyright lobbying effort, 219, 238
  disregard for market surveys, 147
  and Matsushita, 149
  meeting with Sheinberg, 28–29

and Schein, 102–3
and Sony's early years, 38–39
and Strauss, 259–60
succeeds Ibuka, 30, 148–49
and Tapecorder, 43–44
and "time shift," 97
in U.S., 48, 50, 51–52
Universal v. Sony, 34–35, 102, 119, 269
and VHS, 156–57
and videotape recorder, 68
Morita, Kazuaki, 39
Morizono, Masahiko, 63, 64, 90
Morowitz, Arthur, 185, 315
Morsey, Chase, 79
Moses and Singer, 123
Motion Picture Association of America (MPAA), 34, 114, 210–17, 226, 229–31, 232, 233–35, 249–50, 269, 322–23
see also copyrightists; Valenti, Jack
Motion Picture Patents Company, 169–70
Motorola, 74, 76, 77
movies:
copying of rented, 302
illegal distribution of, 171, 321–22
public versus private viewings of movies, 322–23
rental of videocassettes, *see* video rentals and dealers
sale of right to exhibit, 169
sales of videocassettes, 173
Mullin, John, 42
Muntz, Earl, 51
Murase, Tsuzo, 160–62
Murdoch, Rupert, 317

Nader, Ralph, 250
Nagai, Kenzo, 41
Nakagawa, Yasuzo, 45, 64
Nakao, Tetsujiro, 150
National Association of Religious Broadcasters, 119
National Basketball Association, 271
National Broadcasting Company (NBC), 26, 57, 119, 320
National Citizens Committee for Broadcasting, 250
National Collegiate Athletic Association, 119
National Football League, 106
National Hockey League, 106
National Music Publishers' Association, 214, 242
National Rifle Association, 33
NEC, 238–39, 307
Newsweek, 320
New World Cinema, 187
New York Times, 76, 77, 229, 303, 304–5
NHK, 40–41, 61
Nickelodeon Video Showcase, 323
Nicholas Nickleby (Dickens), 125
Nielson survey, 319
Nikkei Business (magazine), 309–10
Nimmer, Melville, 205
Nimmer on Copyright (Nimmer), 205
999 Uses of the Tape Recorder (pamphlet), 44
Nintendo, 275–76
Nishimura, Weston, 201, 236, 315
Nofziger, Lyn, 293
Nolan, Peter, 33
Northcott, Hall, 293, 295

O'Brion, Richard, 70
O'Connor, Sandra Day, 267, 269, 271, 272–73
Ohga, Norio, 151–53, 306–7, 310
Ohsone, Kozo, 311
Okamura, Shiro, 91

PACs, 256–57
Paley, William S., 75, 76, 78–79, 194–95
Panasonic, 140, 163, 165
see also Matsushita Electric Industrial Company
Paramount Pictures, 182, 207, 285, 316, 328
Parliament, 123–24
Parris, Stanford, 203, 206, 212, 218, 219, 220, 224
Pasanoff, William, 129–30
Pascale, Richard Tanner, 140
Patterson, Edward, 250

Patton, Boggs, and Blow, 239, 240
Pet Rock Video, The, 327
Pfizer, 220
Phillips Corporation, 48, 70, 92, 308, 310–11
PHP Institute, 149
Pinocchio (movie), 318
piracy, movie, 171, 321–22
Playboy (magazine), 325
Polaroid, 74
political action committees (PACs), 256–57
Pollack, Roy, 161, 162
Poniatoff, Alexander M., 57
pornography, 183–84, 318
Powell, Lewis, 273
press, the, 248–49
  and rental bill, 302–3
Price, Frank, 207
price-fixing, 289
Principles of Scientific Management, The (Taylor), 137
Publisher's Weekly, 126
PV-100 (Sony VTR), 67

Quasar, 140, 141
  see also Matsushita Electric Industrial Company
QUBE, 112–13

Radio Corporation of America (RCA), 47, 54–56, 60, 74, 79, 81, 88, 160–63, 164, 165, 220
radios, transistor, 49–51
Railsback, Thomas, 226, 249, 250
Rambo (movie), 316
Reagan, Ronald, 26, 100
Reagan administration, and rental bill, 293–94
Recording Industry Association of America, 214, 247
Regency, 50
Rehme, Robert, 207
Rehnquist, William, 242, 244, 272–73
rentals, movie, 170–71
Reynolds, Nancy, 232
Roberts, Steven, 172, 179, 193, 207
Roberts, William, 64–66, 70, 80
robots, 141
Rodino, Peter, 255
Rogers, Fred, 106, 119, 270, 271
Roizen, Joseph, 59, 63
Romani, Romano, 288–89, 294–95, 298, 299
Rosenfeld, Meyer, and Susman, 23, 28, 32, 100
  see also Kroft, Stephen
Rosenfelt, Frank, 207
Rosenman, Colin, Freund, Lewis, and Cohen, 99–100, 102
Rosiny, Edward, 99, 100
Ross, Steven, 188, 194
Rothenberg, Stanley, 123, 135
royalty, *see* copyright
Rubenstein, David, 239–40, 249, 250, 290, 292, 296, 298–99
Ruid, Paul, 32, 34, 117
Ryan, Thomas, 117–18

Sanyo, 92, 150, 157, 158–59, 238–39, 307
Sarnoff, David (General), 54–55, 163
Sarnoff, Robert, 79
satellite dishes, 322
Satoh, Masaaki, 157, 167, 306
Sayco Doll Corporation, 127
Schein, Harvey, 28–30, 34–36, 53, 96, 97, 98, 102–3, 119
schools:
  tape recorders in, 44, 45
  VTRs in, 69–70
Schroeder, Pat, 226, 251, 255
Schwartz, Robert, 239
Screen Actors Guild, 26
Sears, Roebuck, 82, 83–84, 85, 220, 237
SelectaVision, 79, 88, 164, 165
7-Eleven stores, 315
Shales, Tom, 204, 313, 319
Shapiro, Gary, 221–22, 235–37, 240, 247, 271, 289
Sharp, 157, 158, 159, 238–39, 302, 309
Shaw, Pittman, Potts, and Trowbridge, 239

Sheinberg, Sidney:
  backround of, 24–25
  debate with Schein, 98
  Donkey Kong case, 275–76
  and Doyle Dane letter, 21–22, 24
  on film industry, 34
  and MCA, 27–28
  meets with Morita, 28–29
  meets with Schein, 28–29, 35–36
  on movie piracy, 171
  on royalty solution, 205–6
  after Supreme Court decision, 284
  and Universal pictures, 27
  and *Universal v. Sony,* 108, 116–17, 274–75
  on VCR's impact, 323
Shibaden, 66
Shiraishi, Yuma, 144, 145
Signature V (VTR), 65
Sills, Beverly, 226–27
skip-field recording, 67–68, 71
  and Cartrivision, 82
Slayman, Gary, 296
Smith, Rob, 232, 233
Snape, Dale, 232–33, 235, 240, 245, 258–59, 261, 269, 297, 300–301, 302
Sneeden, Emory, 250, 259
sokaiya, 306
Solomon, Risa, 291–92, 296, 299
Sony:
  and American Film Institute, 275
  Ampex suit, 68–69, 100
  annual shareholders' meetings, 306–7
  Betamax, *see* Betamax
  and Betamax compatibility, 94–95, 143, 146
  camcorders, 308–9
  collaboration with Ampex, 63–66
  competition with Matsushita, 142–43
  eight millimeter strategy, 307–8
  Emmy Award, 265
  and Home Recording Rights Coalition, 224, 237–38, 289–90
  Ibuka's aspirations for, 39–40
  image problem, 304–7, 310
  market research, 51
  market share for Betamax, 165, 304, 306
  mini compact disc player, 310–11
  name, 39, 49, 50, 53
  and Phillips, 310–11
  research and development, 89–90, 305
  tape recorder, 41–45
  transister radio, 49–51
  Tummy Television, 51
  U-loading VCR, 71–72
  U-matic, 21, 72, 89–91, 142–43
  in United States, 51–53
  videotape recorder (VTR), 62, 66, 67–70
Sony Corporation of America (Sonam), 28
  counsel in *Universal v. Sony,* 99–105
  Nine West, 34
  reaction to Universal's suit, 34–35
  and Schein, 30, 102–3
Sony's Counterattack (Mimura), 307
Southland Corporation, 315
Speak Up with Confidence: How to Prepare, Learn, and Deliver Effective Speeches (Valenti), 262, 328
Spectator, 126
Spiegel, Dan, 212, 213
Squires, Sanders, and Dempsey, 220, 221
Stanton, Frank (Cartrivision), 81, 82–83, 86, 87, 88
Stanton, Frank (CBS), 76
staple article of commerce doctrine, 120, 134, 270
Stationers' Company, 123
Statute of Anne, 124
Stein, Jules, 25–26
Sternman, Joel, 110
Stevens, John Paul, 267, 269, 270, 271, 272
Sting, The (film), 27
Stowe, Harriet Beecher, 127
Strauss, Robert, 211–12, 249, 250, 259–60
studios, *see names of particular studios, e.q.* Universal Pictures
Stumpf, Richard J., 119
Sugaya, Hiroshi, 92, 94–95, 160, 161
"SuperBeta" VCR, 308
SV-201 (Sony VTR), 66
Sweet, Robert, 275–76

Swidler, Berlin, and Strelow, 249
Sylvania, 165
Synar, Michael, 256

Takano, Shizuo, 92, 144, 309–10
Tamiya, Kenji, 219, 259, 269
Tanaka, William, 239
Tannii, Akio, 162
Tapecorder, 43–45
tape recorder, development of, 41–45, 57
Targeted Communications, 232–35
Tatum, Donn B., 112–13, 207
Taylor, Frederick W., 137
technology:
  and copyright law, 127–32, 243, 277–83
  displacement of old by new, 318–19
television broadcasting:
  networks, 319
  and videotape recording, 55–56
Television Digest (trade newsletter), 34, 80, 81, 84, 85, 86, 87, 294, 309
Texas Instruments, 50
Thorn-EMI-HBO Video, 316
3M, 220, 237
Thurmond, Strom, 213, 215, 260
Tigert, Ricki Rhodarmer, 231
Time (magazine), 317
Time-Life Films, 324
time-shift and time shifting, 97, 107, 230, 271
Toamco, 66–67
Tokumitsu, Hirobumi, 145, 151–53, 154
Tokyo, 37–38
Tokyo Tsushin Kenkyujyo (Tokyo Telecommunications Research Institute), 38
Tokyo Tsushin Kogyo (Totsuko), 39–41
  see also Sony
Toshiba, 47, 62, 66, 72, 92, 150, 157, 158–59, 238–39, 307, 309
Townsend, Robert, 178
transistors, 46–51
transverse scanning, 59, 61, 73
Tribe, Laurence, 231, 249, 253
Trinitron television, 93, 95, 143
TR-63 radio, 50
Trust, the, 169–70
Tsunoda, Koichi, 69, 70, 72
Tummy Television, 51
TV Digest, see Television Digest
TV Guide, 174
Twain, Mark, 126
Twentieth Century-Fox, 76–77, 171–73, 175, 193, 195, 206–7, 285, 287–88, 293, 317, 325
Twentieth Century-Fox Telecommunications, 172

U-Haul Company, 315
U-loading VCR, 71–72, 154–55, 159, 167
U-matic VCR, 21, 72, 73, 89–91, 144–45, 265
United Artists, 188, 190, 207
U.S. Air Force, 75
U.S. Congress:
  copyright revision (1976), 23, 104, 131, 254, 280, 281
  extension of copyright to films, 128
  first copyright law, 126–27
  and grassroots campaigns, 292
  hearings, 247–48
  intent of copyright law, 104–5
  Ninety-ninth, 302
  and rental bill, 285–300
  support for royalty law, 224–25
  and Supreme Court decision, 269–70
  and technological change, 280–81
  and videorecording, 203–5
  see also U.S. House of Representatives; U.S. Senate
U.S. Constitution, 124
U.S. District Court for the Central District of California, 108
  see also Universal v. Sony
U.S. House of Representatives
  hearings, 225–26, 286
  Judiciary Committee, 212, 253–57, 281–82
  MPAA lobbying in, 215–16, 226–27
  PAC contributions to members of, 256–57
  see also U.S. Congress
U.S. Justice Department, 26

U.S. Office of Technology Assessment, 280
U.S. Senate:
hearings, 226–27, 248, 250, 257–58, 299
Judiciary Committee, 258
MPAA lobbying in, 213–15
see also U.S. Congress
U.S. Supreme Court:
decision in *Universal v. Sony,* 268–69, 272–74
effect on Congress, 258
first oral arguments in *Universal v. Sony,* 241–44
grants certiorari in *Universal v. Sony,* 240
reargument of *Universal v. Sony,* 264–68
restores *Universal v. Sony,* to calendar, 246
TelePrompTer case, 118
vigil at public information office, 244–46, 268
Universal City Studios, 27
Universal Pictures:
conservatism of, 25
and Donkey Kong, 275–76
films, 27
home video, 206
remake of *King Kong,* 22
under Sheinberg, 27
and Sony products, 110, 119
takeover by MCA, 25
television, 27
Universal v. Sony:
amicus curiae briefs, 34, 241
appeal, 132–35, 205
decision in District Court case, 121, 131, 132
defendants, 34
discovery process, 109–11
legal arguments, 104–6, 113–20
nominal defendant, 32
popular reaction to appelate decision, 203–4
U.S. Court of Appeals case, 132–35, 205
U.S. District Court case, 109–21
U.S. Supreme Court case, 240, 241–45, 264–69, 270–74
value to Betamax marketing, 98
and Walt Disney Productions, 33–34
University of California at Los Angeles, Law School, 225
Uno, Sousuke, 260

Valenti, Jack, 207–12
book on public speaking, 262
and copying of rented movies, 302
debates, 247, 262–63
on grass roots, 232, 253
lobbying effort, 211–17, 226, 248, 249, 257–58, 260–61, 285–86, 302–3
and Morita, 259–60
and movie piracy, 322
and reaction to *Universal v. Sony,* 204–5
reference to VCRs, 238
royalty solution to home video, 205–7, 257–59
and Supreme Court decision, 269–70
after Supreme Court decision, 284
testimony in *Universal v. Sony,* 114–15
at VSDA annual meeting, 328
Variety, 261–62, 286, 313, 321
see also Daily Variety
V-Cord, 92
Vestron Video, 324, 326
Viacom, 174
Victor Company of Japan, *see* JVC
video, home:
growth of, 285, 312–28
predictions about, 74, 76
video cameras, 73
camcorder, 308–9
videocassettes and videocassette recorders (VCRs):
and cable TV, 313–14
half-flanged cassette, 148–49
and illegal distribution of movies, 171, 321–22
M-loading, 151
prices of, 314–15
problems with, 70–71, 312–13
sales of, 290, 313
standards for, 72, 94–95
U-loading, 70–71

Video Club of America, 173–74
  see also Blay, Andre
Video 8 camcorder, 308
video home system (VHS), 151–55
  camcorders for, 309
  market share, 167
  and RCA, 160–63
  see also Matsushita Electric Industrial Company; Victor Company of Japan
Videophile, The, 106–7, 166
video recording and recorders:
  and Ampex, 57, 58–60
  cartridge, 71
  color, 70–71
  first prototype, 56
  helical scanning, 62
  FM carrier signal for, 58–59
  principles of, 55–56
  problems, 56–59
  Sarnoff calls for, 54–55
  skip-field recording, 67–68, 71
  and videocassettes, 70
video rentals and dealers:
  and Atkinson, 177–80
  Cartrivision, 80–88
  CBS / Fox Video, 195
  and doctrine of first sale, 179, 290
  franchises, 180
  and grass-roots campaign, 235–37, 240, 290–92, 295–98
  growth of, 181–86, 285, 315–28
  pornography, 183–84, 318
  profitability of, 191
  Warner Home Video, 188–202
Video Retailers Association (VRA), 199, 237
Video Shack, 185
Video Software Dealers Association (VSDA), 199, 237, 291, 297, 326–27
Video Station, *see* Atkinson, George
videotape, advantages over film, 73
Video Week, 133, 194, 300
Vidicraft, 302
Vincent, Francis, 207
VX-100, 148
VX-2000, 150, 151, 153, 159

Wada, Sadami ("Chris"), 219, 220, 237, 259
Waldie, Jerome, 250
Walkman, 305
Wall Street Journal, 52, 70, 81, 82, 137, 214, 303, 311
Walt Disney Productions, 33–34, 110, 112–13, 119, 182, 192, 207, 274–75, 317–18, 328
Warner Brothers, 34, 181, 187, 207, 210, 285, 287–88
Warner Communications, 188, 249–50
Warner Home Video, 187–95, 196–202
Washington Post, 204, 245, 298, 300, 303
Wasserman, Lew, 26–27, 28–29, 110, 113–14
Wayman, Jack, 220–22, 226, 236, 247, 252, 261, 262–63, 288
  and rental bill, 288
Waz, Joseph, 250
Weingarten, Frederick, 279–80
Wells, Frank, 188, 189–90, 207
Western Electric, 47, 48, 49
Wexler, Ann, 232, 249, 250, 253
Wexler, David, 32
Wexler and Associates, 232, 235
  see also Snape, Dale
White, Byron, 242, 267, 269, 273
White, Leonard, 316–17
Whiteside, Major, 40–41
Whitman, Walt, 126
Whittier, John Greenleaf, 126
Wielage, Marc, 107
Wilkofsky Green Associates, 314–15
Williams and Wilkins, 130–31, 272, 273
Willis, Edwin, 254
Winnick, Arlene, 327–28

Yamashita, Toshihiko, 162–63, 261
Yoshiko, Morozumi, 62
Yoshiyama, Hirokichi, 158

Zenith, 49, 74, 159, 160, 165, 307
Zeny, Robert, 322